THE
SCHLUMBERGER
ADVENTURE

Paris mardi 29 Juin 1915
et 30.

Mon cher Conrad,

Je viens de recevoir de Mai
Marti la note pour les répon
aux objections des États Unis —
même temps il me donne l'av
que le brevet relatif à la 2e
demande D 325 a été accordée.
J'ai payé la taxe de délivrance
frs 125 et les frais pour répons
frs 20.0.

Il me semble encourageant
que ce brevet américain ait
été accordé et je t'en félicite
de tout cœur.

Je reste persuadé
sont importante
vue scientifique
... il te ...
champ d'ac...
mettre en ...
Par ...

pas, ...
taxe ...
et ...
breve
nos
...
...

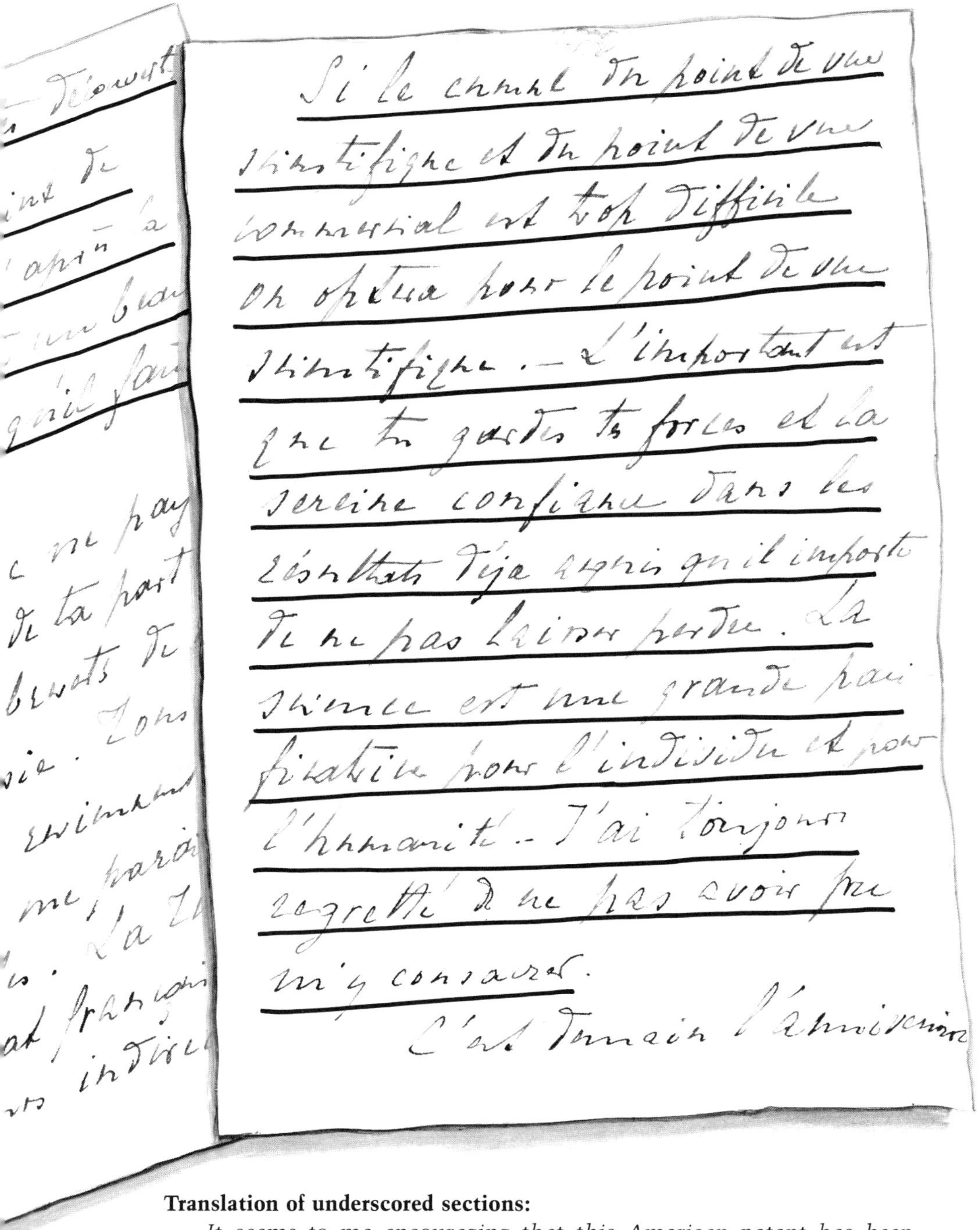

[Handwritten letter in French — partially legible]

Si le chemin du point de vue
scientifique et du point de vue
commercial est trop difficile
on optera pour le point de vue
scientifique. — L'important est
que tu gardes tes forces et la
sereine confiance dans les
résultats déjà acquis qu'il importe
de ne pas laisser perdre. La
science est une grande pacification pour l'individu et pour
l'humanité. J'ai toujours
regretté de ne pas avoir pu
m'y consacrer.
C'est demain l'anniversaire

Translation of underscored sections:

It seems to me encouraging that this American patent has been granted, and I congratulate you with all my heart.

I remain convinced that your discoveries are important from a scientific point of view and that after the war there will remain to you a wide field of activity which will have to be developed.

If the convergence of the scientific and commercial viewpoints is too difficult, it is better to opt for the viewpoint of science. What is essential is that you keep your strength and your serene confidence in the results already acquired and which you must not let go. Science is a great force for peace, for the individual as well as for humanity. I have always been sorry not to have been able to devote myself to this field.

THE
SCHLUMBERGER
ADVENTURE

Anne Gruner Schlumberger

ARCO PUBLISHING, INC.
NEW YORK

All photographs in this book are from the private collection of the author.

Originally published in French under the title *La Boîte Magique*, © Librairie Arthème Fayard, 1977. Translated by Anne Gruner Schlumberger and William Granger Ryan.

Published by Arco Publishing, Inc.
215 Park Avenue South, New York, N.Y. 10003

a Prentice-Hall company

Prentice-Hall International, Inc., London • *Prentice-Hall of Australia, Pty, Ltd.*, North Sydney
Prentice-Hall of Canada, Ltd., Toronto • *Prentice-Hall of India Private Ltd.*, New Delhi
Prentice-Hall of Japan, Inc., Tokyo • *Prentice-Hall of Southeast Asia Pte. Ltd.*, Singapore
Whitehall Books Limited, Wellington, New Zealand

Library of Congress Cataloging in Publication Data

Gruner Schlumberger, Anne.
  The Schlumberger adventure.

  Translation of: La boîte magique.
  Includes index.
  1. Prospecting.   2. Petroleum.   I. Title.
TN271.P4G7813   1982   622'.1828   82-8830
ISBN 0-668-05644-4   AACR2

Printed in the United States of America

10   9   8   7   6   5   4   3   2   1

to the memory of
Conrad and Marcel Schlumberger

Contents

Preface   *xi*

Foreword   *xiii*

Foreign and U.S. Currency Equivalents   *xvii*

1   A Father Backs Science and His Sons   1

2   First Steps   11

3   Discovery of the Alsatian Salt Dome   17

4   A Disappointing Visit to the United States   21

5   An "Eye" to Look Below the Ground   25

6   Blue Pins on a World Map   33

7   Westward Ho!   41

8   A Contract with the Soviet Union   53

9   Eastward Ho!   61

10   Development and Decline in the U.S.S.R.   71

11   Electrical "Logging": From Maracaibo to Burma   81

12   The "Boom" in the United States   93

13   Counterclaims and Compromise   105

14   The War Years   111

15   The International Structure of Schlumberger   119

16   New Enterprises   127

17   From the Magic Box to Aladdin's Lamp   133

Appendix: Schlumberger Limited Today   137

Index   145

Preface

THE scientific development of the Schlumberger technology for petroleum prospecting—so vitally important in the current crisis over energy resources—is brilliantly described by Louis Allaud and Maurice Martin in *Schlumberger, Histoire d'une Technique.** My purpose in writing the present book is to give the "human side" of the story of the men whose achievements spread the Schlumberger method to the four corners of the globe.

"Wherever the Drill goes, Schlumberger goes. . . ." "First in the Field, First in Research.": These are the mottoes of the Schlumberger Corporation. It took the vision and inventive genius of Conrad Schlumberger, his brother Marcel, and Henri Doll, and the courage and initiative of engineers working in jungles and deserts, in tropical heat and arctic cold, under conditions always difficult and often hazardous, to bring these *ideas* to fulfillment.

The pages that follow tell of this worldwide adventure, whose end is nowhere in sight.

A. G. S.

*Paris: Berger-Levrault, 1976; English translation, *Schlumberger, The History of a Technique*, New York: Wiley, 1977.

Foreword

The names of the Schlumberger brothers, Conrad and Marcel, are not well known outside the science of geology and the petroleum industry. Yet their accomplishments, in keeping with a long line of distinguished French scientists, have come to touch the lives of everyone in the world. In time, perhaps, history will come to understand and appreciate the influence these two remarkable men exerted on world events and the world economy through the application of their unique genius.

They could only be thought of as a team. Conrad, physicist, idealist, dreamer, man of ideas, and Marcel, engineer, pragmatist, inventor, man of action, complemented each other perfectly. Together they were able to orchestrate time, talent, and opportunity to put science to work for mankind. In the interest of identifying the hiding places of the world's storehouse of minerals, they devised ways to measure the earth's interior parameters in new and effective ways. Their discoveries and inventions made possible and practical the modern industry that now explores for and produces petroleum. They wrought their miracle at a time when petroleum was coming into its own as the key ingredient of economic progress and when the extent of its existence in the earth was little understood. To a significant degree, the Schlumberger brothers and their associates made possible a petroleum industry of a size and scope beyond the imagining of earlier generations. In fact, it is fair to say that much of the world's oil and gas reserves have been identified by methods the Schlumbergers pioneered.

Likewise — and not incidentally — their work revolutionized the science of subsurface geology. Their findings, the fruit of their technology, resulted in a quantum leap in scientific understanding of how the earth is constructed and composed. The same technology that has explored the earth's crust has since given rise to other technologies that have helped to make possible the exploration of space.

For these reasons, it is timely that Annette Gruner Schlumberger has given us this history of the origins of Schlumberger Limited, the worldwide organization founded by her father and uncle. *The Schlumberger Adventure* was originally published in France as *La Boîte*

Magique (The Magic Box). The name came from the black case that housed an instrument called a potentiometer, used by the Schlumbergers in making geophysical measurements. This book is a biography of an enterprise.

From her childhood days, Mrs. Gruner Schlumberger was intimately associated with the *Société de Prospection Electrique*, parent company to Schlumberger Limited. As a witness to Father Conrad's experiments in the basement of the Ecole des Mines in Paris, as aide to her father and Uncle Marcel, and as the wife of Henri Georges Doll, who headed the company's research and engineering departments and served as chairman of the board of directors, she was personally involved. The great-great-granddaughter of François Guizot, premier of France under Louis Philippe, she also brings a sense of history and an understanding of the international politics that shaped post-World-War-I Europe to her story of the company's growth.

On one level, *The Schlumberger Adventure* is about geophysics and petroleum engineering, subjects explored in detail by Louis Allaud and Maurice Martin in their book *Schlumberger—The History of a Technique*, sponsored by Mrs. Gruner Schlumberger. But *The Schlumberger Adventure* is not a scientific treatise of equations and technical data. It is much more a story of living people, of human relationships, of personal ambition, of life's triumphs and disappointments. It is a behind-the-scenes account of individual dedication and group achievement. It is a spellbinding tale of personal heroism, courage, and sacrifice by those who associated themselves with the enterprise in its earliest days. For along with technical genius, Conrad and Marcel Schlumberger possessed qualities of leadership that inspired others to follow gladly and meet willingly the most awesome challenges in the interest of their cause.

In the face of technical obstacles, financial loss, political problems, world depression, world war, professional hostility, disinterest and disbelief, and legal challenges, the Schlumbergers and their talented engineers marched toward their goal. And they reached it. Death took Conrad in 1936, well before his company earned significant financial success but not before he knew that it was inevitable. Marcel died in 1953, leaving behind an organization that has become today a multi-billion-dollar corporation doing business in more than 75 countries and engaged in providing a full complement of oilfield services and the manufacture of measurement and control instrumentation and computer components. And there Mrs. Gruner Schlumberger ends her story.

The story of Schlumberger does not end, to be sure. The firm that Conrad and Marcel Schlumberger launched continues on course. *The Schlumberger Adventure* gives us a valuable look at a remarkable record of achievement by a truly gifted family and their equally gifted associates. No one could be more grateful to her for this history than those of us who were allied with the Schlumbergers in those early days and whose lives were shaped by the strength of their characters.

For a labor of love performed in the Schlumberger tradition of excellence, thank you, Annette Gruner Schlumberger, from us all.

W. J. GILLINGHAM
former Executive Vice President
Schlumberger Limited

Foreign and U.S. Currency Equivalents

	French Francs	French Francs 1981	U.S. Dollars
1914:	500,000.00	4,145,000.00	690,000.00
1926:	500,000.00	675,000.00	110,000.00
1930:	5,000,000.00	7,450,000.00	1,250,000.00
1934:	2,000.00	4,000.00	700.00
1936:	400,000.00	772,000.00	130,000.00

	Rubles		French Francs	French Francs 1981	U.S. Dollars
1932:	40,000.00	**1932:**	155,000.00	283,000.00	50,000.00
	22,000.00		86,000.00	156,000.00	26,000.00
1936:	400,000.00				
before October 26		**1936:**	1,200,000.00	2,316,000.00	390,000.00
after October 26			1,700,000.00	3,281,000.00	550,000.00

Until November 15, 1935:   1 ruble = 3.89 French Francs

From November 15, 1935
to October 26, 1936:   1 ruble = 3.00 French Francs

After October 26, 1936:   1 ruble = 4.25 French Francs

THE
SCHLUMBERGER
ADVENTURE

1

A Father Backs Science and His Sons

THE scene is the basement of the Ecole des Mines* in Paris, the time from 1911 to the outbreak of World War I in 1914.

My father, Conrad Schlumberger, Professor of Physics at the Ecole des Mines, bends over wooden crates filled with sand. Wires and more wires run out between the crates' panels into a black box. My father is listening through earphones connected to this box. What is he listening to?

Often on Sunday, my sister Dominique and I played around show-cases displaying labeled rocks in the chilly galleries of the Ecole des Mines. Even when I hoisted myself on tiptoe I could barely see the stones. "You're too little, you're too little," my sister chanted mock-ingly, though I was older than she was. (True: little then and little now.)

Annoyed, I would run to watch what my father was doing. There, anyway, I was tall enough to see all that was going on. I was fascinated by the earphones and electric wires that came out of a crate and went into a bathtub filled with water where two sticks—also connected to wires—floated. I was sure that my father was a magician. Had I not seen him bending over a black box and finding coins buried in sand or clay? I must have been seven years old when I learned that this box, which guided him to hidden treasures, bore the mysterious name "potentiometer."

My mother silently viewed the scene, seated on whatever she could find to sit on. Her legs were crossed, and her right foot made a rotary movement like a prayer wheel. We knew that after a large num-

*Graduate Institute of Mining Engineering.

1

ber of revolutions it would be time to go home. "Yes, yes, Louise, let's go," my father would say, "but I haven't finished." On the way home, my father, my hand firmly held in his, answered my questions. Why had he taken our baby bathtub? — Because it is copper. Why did he always say that he hadn't finished? — Because every piece of work, however modest, requires lots of time and patience. "How much?" I asked. — How much time and patience? As much as it takes to finish the work and finish it well.

This did not sound logical to me. If every piece of work had to be brought to its end, why didn't he finish his? I did not known that some kinds of work are never finished and that what he was working on would take more than a lifetime.

Years later, my father wrote that the professorship he had earned at the Ecole des Mines in Paris fortunately provided the long vacations needed to think through the possibilities of applying physics to mining in general, and to prospecting in particular. So in 1912, Conrad started to follow the relatively untrodden paths of electrical prospecting — by which he meant the exploration of the earth's subsurface by electrical processes applied on the surface.

Electrical prospecting is one of those mixed studies which are based on very diverse ideas. It is neither fish nor fowl, displeasing researchers who prudently specialize in a traditional compartment of science. Indeed, if one is to tackle the problem, one must be a mathematician-physicist-geologist-engineer with a taste for experimentation and the open air. Elementary knowledge is perfectly sufficient — but one must either have such knowledge or be eager to acquire it; this doesn't happen as often as one might think, since the field has not yet been widely studied, despite the importance of the prospects it offers. "I am determined to be a technician who goes beyond the study of pebbles,"* Conrad said.

I have before me a reproduction of a sheet of heavy drawing paper, two thirds of which is covered by roughly circular lines. The lines are nearly concentric and they spread out, draw together, or break off here and there. Complete or partial outlines show the location of the château, the garage, lawns, ponds, farm, farm buildings, and so on. The lower third of the sheet is covered with fine, close writing slanting to the right. There are no erasures. The text, which is long, minutely detailed, and divided into five paragraphs, describes the results of a series of experiments carried out on the grounds. In the upper left cor-

Le Puits Qui Parle, 1921.

ner is the note: "Map of the equipotential curves traced by direct current. August–September 1912. Scale 1/1000." My father's love of research had impelled him to make his first full-scale measurements on the family estate of Val-Richer in Normandy.

The lesson he drew from this effort seemed clear to him: The grains that made up the rocks might act as insulators, but the electrical conductivity of the rocks was in proportion to the greater or lesser degree of salinity in the water that impregnated them. Therefore, the map of potentials could show significant contrasts even when the subsurface contained no conductive deposits.

At intervals between the autumn of 1912 and August 1914, a number of studies were made in the iron-rich basin of Normandy, in the Departments of the Gard and Rhône, and also in Spain. These studies confirmed Conrad's idea that electrical prospecting could cover a much broader field of research than the location of metalliferous deposits.

At Fierville-la-Campagne (Calvados), Conrad determined that the two blocks of terrain that were separated by a fault connecting strata of different composition (e.g., schist and limestone) showed variations in electrical field at the level of the break between them. In 1913, at Soumont (in the same basin), while he was determining the trace of a fault, he succeeded in following a vein of ore using a procedure that extended beyond the range of older methods. This constituted a concrete application of Conrad's method to mine prospecting. In April of the same year, at Sain-Bel (Rhône), electrodes planted in the soil above a stratum of pyrite buried 100 meters deep brought to light differences of potential at times when no current was sent into the soil.

Repeated over different terrains and confirmed in the laboratory, this observation supported the hypothesis that an ore containing pyrite acts like an electrical battery. The discovery of this phenomenon, called *spontaneous polarization*, enabled my father to draw up a chart of equipotentials which threw into relief not only the Sain-Bel lode already being mined, but also the contour of a deposit unknown until then.

Some months later, at the copper mines at Bor (in Serbia, now eastern Yugoslavia), the newborn geophysics scored again when spontaneous polarization was used to reveal a deposit of considerable magnitude. Dominique de Ménil, my sister, tells me that these studies, which Conrad continued to regard as experimental tests, were in reality a genuine method of mineral prospecting—a method which astonished observers by its exactness and rapidity.

This penchant for research was not entirely a matter of choice. Paul Schlumberger, my grandfather, a man with an imaginative turn of mind, would have liked to devote himself to science; it was therefore natural for him to push his sons in that direction. After the Franco-Prussian War (1870–1871), when he was twenty-two, Paul had his life ahead of him. Would he stay in Guebwiller and take over the family textile factory? Did he want to live in Alsace now that it was Prussian? He did not like the new atmosphere.

> *We get news of France through the Swiss newspapers. Confronted with the disorders that have broken out in the industrial cities of France, a lot of people are secretly pleased with the Prussian occupation. Some of them have gone so far as to say that without Prussians the industrialist would be faced with an intolerable situation.**

Paul shrank from emigrating, yet he thought seriously about leaving Alsace. His intention was to go to England, and then, once he had learned English, to America. He would be leaving a family environment which was dear to him but in which he was not always happy.

> *Getting up, eating, exchanging a few meaningless words and talking politics—is that all life is about? Yet that is family life which, in spite of everything, is a high ideal to me. I can be with my mother a whole hour without our exchanging a single word, a single affectionate word, no, nothing. What good is it to love each other if we do not live in communion with one another?***

It was understood between his father and himself that Paul would do a kind of apprenticeship in the textile factories in Liverpool.

> *And from there I would set sail for America! That would be great if commerce and industry suited me, but I don't have the qualifications. I am much more attracted by scientific study. I would like to become a professor. . . . My father rebukes me severely. Why all these reproaches?—Because I'm not earning any money. But why take up a profession for which I have no inclination? That's absurd.****

After many an argument, he went off to Liverpool in May 1871. Full of goodwill, having decided to go into industry, Paul no longer balked at following the path that was marked out for him. He was not

---

*Paul Schlumberger's Diary, March 1871 (private collection).
**Ibid.
***Ibid.

innovative, but he thought clearly and logically, and perhaps had some fitness for leadership. "My stay in England is commercializing me," he wrote after several months in Liverpool. "In the end I will come to like it, but I am not forgetting that I am a man of ideas, not a man of action. My decisions are based on reasoning, not on facts alone."

My grandfather returned to Guebwiller in the autumn of 1873. He noted in his diary:

> *Since my arrival, I have had no disagreeable discussions with my father and I think I can get along with him. I am struck by the number of ideas we have in common. Family is important, it is a base, I am more and more convinced of that. I am thinking of staying in industry as carried on at Nicolas Schlumberger et Compagnie. When the boss cannot run everything by himself, he has to have people in his employ who are well paid and, in one way or another, share in the profits. Public opinion despite the favorable turn of affairs is still hostile. The conquest will become a stern reality the day our sons have to join the army. Tranquil living may bore me, but on the whole it suits me. My one ambition is to love a woman and marry well.**

This ambition was realized in the person of Marguerite de Witt, granddaughter of the historian and statesman François Guizot. Paul met her in Paris in 1876. Five sons and one daughter came from this union.

In the families of textile manufacturers, everybody worked either in the weaving mills or in the construction of weaving machinery. Nicolas Schlumberger, my great-grandfather, had perfected a spinning machine; so, in the lineage of these Alsatian industrialists, the post of *filateur* — proprietor of a textile mill — passed from father to son. It was not until years later that Paul's preference for science found expression in the work of his sons Conrad and Marcel.

In 1900, all of Paul's sons except the youngest were living in France. When his last son left home, my grandfather grew impatient to be with his children and to fulfill the wishes of his wife, who worshipped the memory of her ancestor Guizot. Since the law allowed Alsatians to recover French citizenship, Paul decided to leave Alsace for good and to turn over his interest in the family business to his brothers.

He took up residence in Paris, an *émigré* with capital to invest. He had always felt constrained by his business affairs; now he saw the fruition of his youthful dreams in his sons' scientific gifts. They, at least,

---

*Diary kept daily by Paul Schlumberger.

were venturing into virgin territory rather than following the beaten path. I can still hear my grandfather telling me: "You see, my child, in our family, aptitudes alternate every two generations. There are the inventors and the adapters. I'm not putting either of them above the others: I'm just stating a fact."

The future would bear witness that the two aptitudes were united in Conrad and Marcel. The fact remains, however, that Paul Schlumberger, proud to see in Conrad a turn of mind which to him was second to none, encouraged his son to devote himself to pure science. The younger son, Marcel, was an engineer graduated from Centrale.* Paul recognized in him the spirit of exactness and methodical approach that was perfectly suited to temper the creative flights of his older brother Conrad. Marcel was passionately interested in mechanics and was then working on an automatic gearshift for automobiles. In 1909, after his marriage, he had turned to the mining industry, an occupation that sent him travelling in distant lands — Algeria, Asia Minor, Crimea. "I wish," their father said, "to have my two sons doing research on electrical prospecting." Conrad's speculative orientation and Marcel's work were to converge in the fulfillment of his hopes.

Being opposed to war by nature and conviction, Conrad had joined a society for Franco-German rapprochement. The events of 1914 brought a harsh awakening, which swung him to the side of invaded France. As a captain of artillery Conrad succeeded in pinpointing enemy batteries by measuring the refraction of sound waves. Mintrop, a German, was engaged in the same experiments. When peace was restored these two men found themselves face to face, no longer as enemies but as competitors in geophysics.

The horrors of war left their mark on Conrad. Once free of his military obligations, he wanted to give up his research and engage in social and pacifistic work. In an unpublished essay written during the war entitled "*Nos torts et nos devoirs*,"** he explained and defended his decision. But my grandfather, who had spent all those years waiting for the return of this son, did not leave him the option.

One evening, I learned that a covenant had been signed before a notary on November 12, 1919. I quote it in full; the document helps reveal the character of Conrad's father, with his combination of cunning and naiveté.

*L'Ecole Centrale des Arts et Manufactures.
**"Our faults and our obligations"

I agree to disburse to my sons Conrad and Marcel the funds necessary for research study in view of determining the nature of the subsurface, in amount not exceeding five hundred thousand francs. On their part, my sons will agree not to disperse their efforts, and to abstain from research or inventions in other fields. The field of activity is vast enough to satisfy their inventive genius by its investigation: they must devote themselves to it entirely. The scientific interest in research must take precedence over financial interest. I will be kept informed and will be able to express my opinion as to important directions and expenditures to be made or not to be made. The sums disbursed by me are a contribution on my part to primarily scientific and secondary practical work which I consider to be of the highest value and in which I take an interest. Marcel will bring to Conrad his remarkable competence as an engineer and his common sense. Conrad, for his part, will be the wise physicist. I will support them.

Jean, the oldest brother, noted in his diary for December 2, 1919: "Surprised and moved by the perspicacity with which my father has been able to come to the aid of his sons, freeing Conrad and Marcel, the one from his daydreams and the other from his isolation, by providing them with the means to pursue more confidently their research in mine prospecting."

Talking about Conrad's political ideas, which in my grandfather's eyes were nothing but mirages, Jean Schlumberger continued:

So then, nothing has changed! The same men who had not been able to ward off the catastrophe went right on imposing their obtuse, sterile methods. The infinite suffering, the millions of deaths, had not in the least shaken the habits of ministries and chancelleries; the stunning joy of victory had begotten nothing new or liberating. . . . Among all the men—colleagues, friends—with whom Conrad renewed contact, he met nothing but misunderstanding and scandalized reactions. Putting his immense disappointment out of his thoughts, he applied himself totally and exclusively to his own work. Along with teaching he continued his laboratory experiments and their verification in the field. From that time on, his life was inseparable from that of the progress of his research.

Joined together by their father, the two men became an inseparable team. Question the one without the other and the answer would be, "I'll talk about it with my brother Marcel . . .," "I'll ask my brother Conrad."

As often happens with family affairs, the children were very much aware of these events. I retain a vivid memory of the wisps of conversation that I caught on the fly, so to speak, and translated for my private use. It was as if I had my eye glued to a kaleidoscope, except that now the shimmering play of colors was transformed into a gymnastic of words. Whenever my father sat down at the piano, I hid underneath so as not to miss anything, in case one of those words should fall from his lips between two chords. Even on Sunday when, wearing goatskin coats, we went on picnics (never mind the weather, one had to "take the air"), I took time from play in order to catch a word on the sly. I listened and listened, not understanding a thing, yet what was said around me was captivating.

My father aroused my enthusiasm, but my Uncle Marcel intimidated me. He was tall and handsome. One did not play with him — his bushy eyebrows forbade it. On Sunday, axe in hand, he was off to the woods to chop down trees. Later I realized that he had a surplus of energy to get rid of: His impatience at not achieving immediate and positive results in his work drove him to physical effort. "One can do a lot of things with string," my father said jokingly. Leaning over a drawing board with a cigarette dangling from his lips, my uncle smiled, saying that string had to be untangled, just as precious metal had to be freed from its dross.

Intelligent and plainspoken as they both were, my mother Louise and my aunt Jeanne frequently disagreed. They thought alike only in admiring their respective husbands, each of them supporting her own spouse both with unfailing confidence and with a generous dowry. My grandfather's capital had its sacred side; one dipped into it sparingly. "Careful, it's Father's money," was a ritual saying. In fact, long after its formation in 1926, the Société de Prospection Electrique* continued to function by drawing advances on account agreed to by the two brothers, who did not receive a regular salary before 1935. Until electrical coring in the U.S.S.R., Venezuela, and the United States brought in more substantial returns, the rare receipt of income came from some mining studies and electrical prospecting surveys in France and abroad — resistivity measurements in the Moselle coalfields, tracing subterranean faults in the Rhineland, searches for lignite in southwestern France and for pyrite in Poland, the former Belgian Congo (now Zaire), and Rhodesia (now Zimbabwe). The application of the resistivity method to

---

*The in-house nickname for which was *Pros* (pronounced as in *prospecting*).

prospecting petroleum-bearing structures was not implemented until 1923, in Romania.

My grandfather grumbled in his beard and went along with the game. "You look like a sensible man," he used to say to one of the engineers. "My sons will ruin me; I've already spent 500,000 francs. What do you think about their work? They have their own way of telling me the story . . ." Most frequently, their "way" was to crisscross an area aboard an old, light truck. Conrad would climb up beside the driver, Marcel and an engineer (fresh out of school) would pile into the back with the equipment, their legs dangling. Through fields and woods, over tilled land and ditches, everyone did his share. One planted the pegs, another dragged the cables, the third took the measurements.

Notebook in hand, Conrad made his observations. Things had changed since the days when crates of sand and clay were the "field"; now the weaker voltages forced him to follow the movement of the potentiometer needle with a magnifying glass. Measurements were now incomparably more delicate, methods and apparatus continued to be improved; but conditions on the site — the soil too dry, metal pipes underground, parasite currents, rain falling on the black box, and so on — could and did throw the measurements off. The men would return home covered with mud, dead tired. My grandfather's bad humor would be forgotten, his faith wholly reborn — a faith not grounded on concrete reality but on an abstraction: the infallible genius of his sons. Like me, no doubt, all he retained about their work consisted of such high-sounding terms as *telluric, polarization, equipotentials* — words he invested with magical properties. Grandfather's eyes would gleam with fervor and profound reflection; then, as if he had put his finger on a piece of dazzling evidence, he would proclaim loudly and clearly that, given just one great mining discovery, his name would be known all over the world.

Yes, but to realize it, they still had to convince mining companies that it could be done.

2

First Steps

IN the early 1920s, exploration of the subsurface of the earth was usually done using methods which had changed very little for centuries. Progress had been made in geological research, but it still depended on reconnaissance drilling, which was often haphazard. The idea of surface electrical measurements, which had to be made with a great array of cables, spikes, and other apparatus, had little visual appeal. The procedure was not new, and it failed to inspire confidence. The term *geophysics* was hardly used at all. Many geologists (and the mining industry in general) refused to see any promise in geophysics. Tradition and experience, after all, showed that metallic ore was discovered only by making holes in the earth; the rest was sorcery — divining, in fact.

It must be admitted that there was good reason for this skepticism. All kinds of pendulum-swingers and coiners of pseudoscientific jargon were peddling hazy ideas which a complacent — or credulous — press reported. A certain Dr. Moineau announced a radiocondenser that indicated "with certainty" the point on the surface directly above oil-bearing strata, as well as their depth and thickness. Someone named Regis was the proud inventor of a machine producing "Hertzian waves and cathode rays," by which he determined, from a distance of several hundred kilometers, veins of mineral, bituminous deposits, petrol-producing sources, mineral waters and coal." In 1934, Conrad published a plan to award a considerable cash prize to the first of these gentlemen who, in the presence of a competent jury, could identify the nature of various liquids contained in opaque flasks. Not one of them dared to take the risk, or even to present himself.

Although these elucubrations missed the mark, they still created complications. Explaining theories and methods, describing operational techniques, multiplying demonstrations. . . . All this was not enough;

what also had to be done was to conquer inertia, overcome fixed habits — in a word, to educate. In fact, several studies and field operations had been carried out in Sweden and Germany, and also in the United States, where seismology and gravimetry had registered some unexpectedly successful results. In the U.S. — where, said Conrad, "ideas were coming to the boil" — from 1921 on the two brothers had a knowledgeable scout in the person of Sherwin F. Kelly, a young American graduate of the Ecole des Mines in Paris, who had learned their prospecting methods. Kelly had made successful measurements in the copper mines in Tennessee and Michigan; in the anthracite fields in Pennsylvania; in Canada, where cobalt and auriferous pyrites were mined; and at Noranda (Quebec), where electrical prospecting had led to the discovery of an unexpected deposit. But notwithstanding the successful results of the missions, the horizon was still clouded.

Economy was, therefore, the watchword. My father became his own secretary, typing the correspondence, sharpening his own pencils, and filing his papers. It is true that until 1927 (the year electrical coring was developed) the administrative tasks were not time-consuming. Moreover, even with crews at work in the field, little work was devoted to constructing new equipment. Rather than designing and building equipment, what was important was the precise definition of problems, learning to interpret clearly the meaning of the curves — in brief, to speculate. I saw my father and my uncle spend hours and days over diagrams that were difficult to interpret. There were too many parasitic currents, too many disturbing elements; the task was to understand them and to eliminate their interference. In this situation, an increase of personnel would have been superfluous, and as for employing a full-time geologist, that would have seemed like an uncalled-for expense.

Nevertheless, the need for such an assistant was felt more and more. So, in 1922, a young Swiss geologist named Edouard Poldini was hired as a consultant. A year later, while he was in Romania, Poldini was approached by Jules Meny, president of the Steaua Romana Company. Meny was aware of the Schlumberger operations and wanted to know whether the resistivity method could be used to explore petroliferous structures. A prospector and his equipment were dispatched. Some preliminary measurements looked interesting and a contract was signed — at first to cover experiments, but soon broadened to include the systematic exploration of a very large area.

Some months after the beginning of the campaign, near Aricesti in the district of Prahova, the chart of resistivities made it possible to "see" the contours of a dome in which the high degree of salinity seemed to be characteristic (or so it was thought at the time) of the formations bordering reservoirs of petroleum. This discovery of a gas deposit in the Aricesti dome by the resistivity method marked a turning point in the history of geophysics.

Operations in Romania continued until 1940 and even later, terminating only when foreign enterprises were nationalized in 1948. Numerous missions succeeded each other; there were three in 1925, each with its own nickname. Coste and Laubereaux's crew were called the "cooks crew" because they did their cooking on the work site. Gallois and Delord's were the "iron crew" because they had the best returns. Baron and Jost's group were the "brick eaters" because they were so far from any supply center they had nothing at all to eat. The men on these missions, already familiar with surface electrical prospecting, had to learn to use new techniques and equipment as they were developed in the Paris workshops.

Living and working conditions were primitive. The men set up their base in a village, lived in log huts, washed at a pump, ate polenta. The roads were impossible; to go from one place to another with the equipment, an oxcart was often the only vehicle that could get through the winter mud. Communication with Paris was mediocre, contact with the local people difficult. Despite their traditional hospitality, the country people turned a suspicious eye on these invaders, who came with their bizarre equipment, stared at their fields, and trampled their meager crops.

Here as elsewhere, the prospectors succeeded each other and were not at all alike. There were Léonardon, more restless than an anthill when food is scarce; Henquet, a giant with the airs of a Balkan baron; Mennecier, an inventive mechanic whom a faulty tool would drive to the brink of tears; Autric, so nearsighted that he could not tell a bull from a haystack at ten paces; Jost, a self-educated fellow who gave advice to engineers in their tryout period; Baboin, the ambitious; Breusse, the malcontent; Scheibli, the crazy driver; my brothers-in-law, Jean de Ménil, a specialist in international finance, and Eric Boissonnas, a geophysics enthusiast; Poirault, a talented musician who kept his cello close at hand; Poldini, a high-relief character who, with his wide-brimmed hat, ample black cape, and huge boots that came up above his knees, looked like a Corsican bandit; and twenty others. . . . We now can find them all — or almost all — here and there, in the Soviet Union, in Venezuela, in the United States, and in the Dutch East Indies.

In 1923, just off the Esplanade des Invalides, right in the center of Paris, a former café was used as a workshop. A stock-keeper and a mechanic worked amongst crates, cables, electrodes, and black boxes. In the courtyard, a sort of closet with a heart-shaped cutout opening functioned as a latrine. In the only room on the second floor, Conrad and Marcel faced each other across a double desk. A drawing table littered with charts and diagrams left little room for my father to practice his habit of walking up and down, back and forth, eyes on the floor, thumbs in his vest, thinking aloud. Leaning back in his chair, lighting cigarette after cigarette, Marcel listened. The cash box was empty, the salaries hardly worth talking about, the future unpredictable. To the young engineers, Conrad, smiling, preached ascetic living: "When there's hay in the hayloft, everyone will get his share. Meanwhile . . ."

Meanwhile, the unknown and life in the open air stimulated the future prospectors. To make hay, they had to succeed: Life itself depended on success. Soon after being hired, they went off on their missions, their baggage including electrodes and potentiometers. Left to themselves, with a minimum of training and with equipment that often was ill-adapted to the terrain, they found it a real gamble. Anxiety reigned. Must the surveys be cut short, the expenditures stopped? The French mining industry was in a rut, not open to new ideas, slow to react.

Battle-weary, Marcel and Conrad made a firm decision to look elsewhere. Texas, with possibilities on the scale of its vast expanse, lured the imagination. They asked their nephew, Marc Schlumberger, to go there and scout the territory.

This young man had many of the qualities needed for success but had no technical training. For all his brilliance, charm, and mastery of the English language, Marc was unable to interest the oil men in a prospecting method they had never heard of. He was getting nowhere, but this was not enough to upset him. He wrote long letters advising his uncles to be patient and optimistic—"a pool of petroleum isn't found in a week," "electrical prospecting can't help but carry the day," "the oil companies will come running . . ." The proof, Marc maintained, was provided by one of his lady friends, a clairvoyant who had casually told him that he was soon going to find the very thing he was looking for. And what was that, if not petroleum? Of course, while waiting for new orders from headquarters Marc would not take such a prediction too seriously, but (since he hadn't asked for it in any way) it did make you think, didn't it?

His uncles, whose thinking drew on other criteria and who believed that the proof of the pudding was in the eating, did not care much for Marc's literary flights. Moreover, they had recently made contact with the Royal Dutch Shell Company; they therefore decided that one of them would make the trip to America. In June 1925, Marcel boarded ship for the United States.

Shell had shown an interest in the Schlumberger methods as a consequence of the results obtained in Romania. One spring day in 1925, Conrad had arrived in Pechelbronn, France, with Dr. Mekel, head of the Shell Company's geophysical services in the Hague, who wanted to see the measurements being taken in a borehole. The crew, duly alerted, had taken appropriate action: sand on the access road to the well, a splash of paint on the old truck, equipment cleaned and calibrated. The stage was set. Conrad and Mekel arrived, Deschâtre went through a demonstration of the apparatus, Sauvage proceeded with the measurements and traced the diagrams. "Fine and dandy," said Mekel. "Now, go away. I'm going to try it myself." Conrad, I suppose, did not like being taken for a Dr. Moineau and his engineers for accomplices; but then, in a sense, Mekel's caution was understandable. He was allowed to do what he wanted. It would seem that he was satisfied with the result, because one prospecting contract was drawn up with the Roxana Petroleum Corporation, a subsidiary of Royal Dutch in Texas, followed by a second contract for work in California.

In New York, Kelly and a Shell employee were waiting for Marcel on the dock. They eased him through the formalities of debarkation and customs, arranged for his hotel room, and reserved his seat on the train to Dallas. Marcel made some contacts, had lunch with J. C. van Panthaleon Baron van Eck, the president of Shell, and opened a New York bank account to make things look good.

Cases of equipment awaited Marcel in Texas; upon his arrival there, he proceeded with a few preliminary tests that came out well enough. At Pierce Junction, Blue Ridge, Goose Creek, and other picturesquely named places he planted electrodes, ran profiles, measured the conductivity of the terrain. If a cable broke under tension or a resistance in the electric circuit did not work, he improvised an automatic reel for the cable or a new winding for the circuit. At Bayou Serpent Marcel found what he had hoped for, at Cap Rock he found nothing, and everywhere he waded around in his great boots fighting off the mosquitoes.

Get one single good interpretation and other contracts will follow. Leaving to his nephew Marc the matter of cultivating contacts, and to Gallois and Baron responsibility for a first prospecting crew at Freeport, Texas, Marcel set out on the return journey. He was far from suspecting that the day would come when the fabulous volume of business for which he had just planted a humble seed would make this early expense account appear almost comical:

Pullman, Houston–New York	$83.59
Hotel Pennsylvania, New York	6.05
Taxis	3.60
Tips	10.00
Train Boulogne–Paris	f. 4,25

It was financial ruin!

3

Discovery of the Alsatian Salt Dome

IN the person of my husband, Henri Doll, I gave my father the son he had longed for. Henri joined Conrad and Marcel toward the end of 1925, in the small building where they worked and explored their ideas. He was twenty-two years old, a graduate of the Ecole Polytechnique and still studying at the Ecole des Mines. Henri liked nothing better than tackling and solving problems; it was his nature to assess everything with the tape measure of logic. Could he have found a more exciting challenge than to begin his professional life in the new field in which the two brothers were engaged? To that field he devoted his life.

Meanwhile, the mysterious black box of my childhood, as if it too had grown with the years, bristled with ever more complex accessories. New techniques were constantly being developed, and they, in turn, required the designing and testing of new equipment. The modest machinery the three men had at hand was hardly adequate to meet these demands. To my husband, who was explaining the advantages of an improved lathe, my father once retorted, "But what are you going to *do* with such a contraption?" Indeed, it was both too little and too much: too little, because one more machine tool (however improved it might be) could not make possible all the projects that Conrad, Marcel, and Henri had in mind; and too much, because no matter how conclusive the laboratory tests might be, a prototype could not be put into production until tested under actual field conditions. Hence a vicious circle: An apparatus had to prove itself in its natural environment; the natural environment was obviously the terrain; prospecting terrains required the clients' agreement; the clients, despite the performances at Aricesti and elsewhere, were slow in coming to terms.

Then, all of a sudden, Alsace, native heath of the Schlumbergers, gave the prospectors their first spectacular geological success.

In the spring of 1926, the search for salt uplifts (upon which potassium mining in Alsace was dependent) led to the discovery of the anticline at Mayenheim and, much more important, the dome at Hettenschlag. Prospecting the plain of Alsace cost about 500,000 francs of the period; it covered over 300 square miles and necessitated some 4,000 electrical measurements. Dragging the lines over the vineyards in this plain, between Mulhouse and Colmar, demanded the inexhaustible agility and endurance of the prospectors. "Athletic engineers," the advertisement in the *Journal des Mines* had specified. My husband worked with the crews.

Emotion was at a high pitch; a telephone call to Paris precipitated the arrival of my father. He wanted to verify personally the measurements made at different depths, which always produced the same figures. But there it was! Doubt was no longer possible: The resistivity charts certainly revealed five anticlines and a saline dome, of which the upthrusts, leveling off at about 150 meters from the surface, were strung out over some 40 miles!

Psychologically, this discovery was the end of the dry spell for the two brothers. Electrical methods were proving their validity in the plain of Alsace, where, previously, classical geology had failed to find a single salt dome. Some months later, when borings had confirmed the presence of a potassium deposit, the chart of the Hettenschlag dome, appropriately framed, held the place of honor in Conrad and Marcel's office. On February 13, 1928, a paper read at the Academie des Sciences gave an account of their achievements.

Life went on at the Paris office. Income was so meager as to dictate the strictest economy. Afternoon tea brought some of the personnel together. Stalking up and down the room, munching an apple, head bent forward as if to follow the lines in the floor, Conrad talked. This pacing was probably characteristic of the days when, as a professor at the Ecole des Mines, he lectured to student audiences.

Every once in a while he came to a halt, his eyes suddenly alight with pleasure, as if some passing observation had recalled an unexpected memory. His words were clear, his delivery easy; like his writing, they were without mistakes. I do not mean to imply that he cultivated an oratorical or pedantic effect. On the contrary, his speech was simple, direct, reduced to essentials, and when he did not know the answer to a question he said so as a matter of course. Later on, in the U.S.S.R., surrounded by Russian technicians, he was affable as usual, open to questions and even inviting them, always attentive to the ideas of others in spite of the obstacle of translation. On the other hand, in contract negotiations, his tone changed; he possessed the art of putting forth an argument, pleasing a case, bringing out hidden evidence. But here, over tea and crackers, he held work sessions. The engineers listened, Marcel listened. Between the two brothers a process of osmosis produced a continuous communication. Taken up and turned over again and again, the gist of each problem slowly emerged. These discussions were marked by an insistence upon accuracy; great care was taken in proposing anything as fact unless observations confirmed it.

Sometimes technical questions yielded to economic concerns. Should the brothers resign themselves to accepting rather mediocre contracts that would at least enable them to hold on, to continue their research? After all, they had to improve their techniques and dig into their ideas — in a word, to survive. Then there was the question of patents; in 1927, some forty had been taken out in France and as far away as Australia. In theory, they were unattackable, but there were people who were doing their best to get around them. Must the company wage a war of attrition? Make the infringers pay? In addition, the idea was already taking shape that if the enterprise developed in Europe, it might well become worldwide. These debates, over a cup of tea, were indeed sacred moments, but they were soon to disappear. No matter how small the individual success, the business was bound to grow too big for the family framework.

Yet certain habits persisted. Every day I went to pick up my husband and my father. At a quarter past twelve, Conrad with his coat on, armed with an umbrella, was already on the doorstep, calling, "Marcel, you're keeping everybody from lunch!"

Marcel, working on a design, did not answer. His internal clock was not set to the rhythm of the hours; Clairin, his head draftsman, generally wound up nibbling a sandwich at his desk. Marcel, wearing a white blouse, perched on his stool, bushy of eyebrow, smoking cigarette after cigarette, could never tear himself away from his project. One day, his wife wanted him to get measured for a new suit. "No! Out of the question! I have no time to go out, and it's a nuisance anyway!" One week later, the tailor was taking his measurements while he, sitting tight on his stool, back turned, pencil in hand, and cigarette at his lips, obediently lifted his elbow.

Nightfall took Marcel unaware. "What's this, Clairin, you're still here?" Clairin, solid on his feet and square-shouldered, had no idea that he might be invisible. The next day the scenario began again. Marcel came in with some new ideas and a school notebook full of sketches, and the previous day's projects were back on the drawing table. Patient and tenacious as he was, Marcel did not leave that table except to go to the machine shop, where his ideas took concrete form. Open with his workers and technicians but without undue familiarity, he sat on the end of a workbench; crossing his legs and resting an elbow on his knee, his voice low and his gestures calm, he talked with different people about the work under way. And, provided the particular object — most often a prototype — was following the specifications, Marcel let each man do things his own way. When he finally saw his machine — *his* machine — finished, there was a kind of joy in the way he took possession of it, examining it with almost amorous care. More than once I saw him, surrounded by his crew, crouching beside some new apparatus for half an hour, even an hour; and when he straightened up to his full height, his young assistants (who, out of deference, had crouched along with him) unfolded their limbs as if stricken with rheumatism.

4

A Disappointing Visit to the United States

IN September 1926 it was Conrad's turn to land in New York. After some unproductive contracts with the Gulf Oil Company, he set out for Texas. From Dallas to Houston, from San Antonio to Corpus Christi, spending most of his nights on trains, he visited the oil fields and talked with geologists around the derricks.

One evening, with Charrin, his collaborator from the beginning, Conrad left Laredo in a ramshackle Ford to find Deschâtre, who, all by himself, was the only "crew" in the field. At the time, in this region, the paved roads disappeared once the towns were left behind. The trip to Randado—a place not even marked on the maps—was made over dreadful roads. The headlights shone only when the car was driven in first gear; then the radiator boiled over and burst a hose. It took them ten hours to drive eighty miles.

Deschâtre and his five Mexican assistants lived in a tumbledown shanty, all sleeping in the same room and eating the same canned food. In order to drag the cable through the brush they had to use a compass for direction. Of course, there were insects, but what bothered them more was the cactus, whose insidious thorns had invisible barbs that went through clothing and penetrated flesh. In a letter to my mother, Conrad told how he went about extracting the thorns from the coppery skin of his aides. I like to picture him as attentive and brotherly. He took care of these men of mixed Spanish and Indian blood, whose wild appearance he described.

The country teemed with game. There were rabbits, hares, partridges, and deer in abundance, as well as opossums, badgers, wolves, jaguars, and many rattlesnakes. Conrad, who killed two snakes, brought the rattles back to Paris as trophies. In the midst of all this, cattle and horses roamed freely; cowboys herded them to the water holes twice a day, in the best Western style.

From Texas, Conrad went to Louisiana, where Roxana (a subsidiary of Royal Dutch Shell), was prospecting the marshlands and bayous. Starting from Lake Charles, Conrad undertook to explore Bayou Serpent aboard a flat-bottomed boat with a roof that provided shelter in bad weather. Pierre Baron went along with his assistant. They were heavily loaded down with gear: electrodes, signaling horns, measuring apparatus, and wires by the mile. They looked like mountaineers out of an operetta.

The boat, the lianas that entangled the cables, the water hyacinths, the flora and fauna of the remote bayous, stirred Conrad's imagination. Here, nothing was sterile. The slightest movement signaled unexpected life — he could not imagine that this land, bathed in secret water, could deny him. Besides, had not a Cajun, who, of course, spoke French, talked to him about a blowout that had occurred in this very place ten years ago? Yes indeed, and then the chart of resistivities could not lie: a boring must be made here and there. The Roxana geologists, however, shook their heads. Their science led them to distrust this slightly ridiculous "Frenchy" who proposed to show up the good old torsion balance with his patched cables. "I'm a good sport," Van Horst, one of the directors of Roxana, confided to Léonardon, the delegate of *Pros* in the U.S., "which is to say that I'm not afraid to sink $250,000 in a project which gives me a feeling of confidence; but it also means that I won't risk a dime if the thing looks unproductive." Some months later, the contract with Roxana Petroleum, which had seemed so promising, was terminated.

It seems to me that anyone else but my father would have felt as much discouragement as previously he had felt enthusiasm. On the contrary, his natural impulse was to insist that the geologists were right, even though they had cut the ground from under his feet. He played fair. He admitted that while his methods were scientifically proved, the technical side left much to be desired. The apparatus had to be improved, too. The crews were insufficiently trained, and yet everyone was working as hard as possible. From Bayou Serpent to Laredo, from Frio Clay to Reynosa, from Napoleonville to New Iberia, my father went to field after field. Everywhere he heard nothing but anticlines, faults, profiles, sands, salinities, sulphur . . . and five hundred feet, two thousand feet. By letter and telegram he pressed Paris for batteries, stakes, cables, relay resistances. In Houston, he wanted a permanent reserve of three potentiometers, three topographic kits, twelve miles of cable, two hundred iron and one hundred copper stakes, and a long list of miscellaneous items. His vocabulary was already enriched with American oilmen's terms such as *lease, wildcats, seepage, blowout, gusher* . . .

Aboard the *Argonaut*, a transcontinental train, Conrad was then off to California, where contacts with Shell required his presence. Curious contacts. . . . In the San Joaquin Valley, at a place called Lost Hills — a flat, dry, desert region under a blinding sun — one had to make oneself invisible, so to speak. Anonymity was the rule. The formidable Shell Company pretended to hide behind someone named "Mr. Davis," a geologist, who in turn hid under the pseudonym "Mr. Arnold Barber." A letter on the Roxana letterhead, arriving from Dallas, almost precipitated a storm in Paris: "It is assumed that we are working for ourselves," Conrad informed Marcel, giving him only a post-office box number as a return address. Nevertheless, with his crew — two engineers, three assistants, a draftsman, and two cars loaded with equipment — he ranged over a terrain of some seventy-seven square miles, where perfect climatic conditions, without a trace of humidity, preserved the cables from any possible leak. Conrad was ecstatic over the quality of the light, the conductivity of the soil; even the spiders, which he described as huge and mean, aroused his curiosity.

In 1926, the company counted seventeen engineers. There were eight crews with two engineers each — six crews in America, two in Europe. The seventeenth engineer stood by "in case of accident," as my father said. René Viry and Robert Nisse made up the California crew. Viry had orderliness and common sense to burn. Nisse was not methodical — he was more zealous than efficient — but was an excellent geologist. Unlike as they were, the two got along famously. Conrad stayed with them for three long weeks, sharing their life and working as hard as they did. He took measurements, discussed the results, bought a typewriter, dry batteries, and cables in Los Angeles and, just as Marcel had done before him, kept an account of the smallest expenditures. He was very hopeful. With the seismograph and the gravimetric balance he had been investigating this area for two years without any result; nevertheless, he concluded that he and his electrical method could remain on the scene for some time to come. After all, wildcatting cost fortunes, while his crew . . . well, it was cheap!

On November 1, 1926, on his return journey, Conrad stopped in Dallas and Houston to renew contact with Gulf Oil and claim what was due to him: "a bagatelle of $7,500 with which to refurbish our account," he wrote to Marcel. On the 17th, he boarded the *Aquitania* in New York. He had had so much to do, so many things to get off the ground, that he had overstayed his leave; now, if only his wife did not hold it against him . . .

Proud as he was of his sons, my grandfather hoped that, once their success was assured, he would be free to wander in his beloved woods. So that, after my grandmother's death, his weary spirit abandoned itself happily to nature, and he frequently took long walks in the forest. Wearing his cherished green Austrian cape, his white hair floating from under the brim of an old felt hat, he looked to me like an ancient shepherd. Often, he would not come back from these hikes until dark, returning home in bad humor. Was he then to be scolded like a mischievous little boy? Hadn't he given his children enough money, enough support? Why should "they" bother him when even the animals were his friends?

Sometimes, he allowed me to accompany him into the forest. He would point out roots or flowers with his cane, angered when I did not know their names. One day, deep in the woods, he said to me, as if he were reluctantly unburdening himself of a grave secret: "Annette, you're not like the rest of them, are you? When I'm dead, will you see to it that they at least leave me my cape? You must tell them. . . ." I did, And today, more than half a century later, his cape hangs on a hook behind the Alsatian stove at Val-Richer, waiting for him.

The next day, October 15, 1926, was a Norman day like any other. It was drizzling. Suddenly, the family home fell silent. My grandfather had joined the earth he loved. The old man's fantasies were laid to rest.

5

An "Eye" to Look
Below the Ground

U NTIL 1927 the investigation of the subsurface had been confined
to measuring the resistivity of terrain through electrodes planted
in the soil. The ore (solid or liquid) was farther down. Like blind men
feeling their way, wells and borings could skirt the deposit being sought
without "seeing" it, although at times it was very close. What was
needed was to replace the stakes on the surface with sondes* at the
bottom of the holes so as to explore their walls. This idea, like all fruitful ideas, grew out of an earlier intuition, not yet formulated but awaiting its time. Taking measurements in boreholes was a logical and technical extension of surface electrical prospecting. My husband tackled
the problem head-on.

In August 1927 Henri, impatient to start making measurements in
boreholes, settled in Péchelbronn, the only oil field known in France at
that time. In fact, to call it an "oil field" is saying a lot. As the name
Péchelbronn indicates, it was originally a spring of pitch or tar. It seems
that in the Middle Ages, pitch was used as a salve or liniment against
a variety of diseases. Its supposed curative powers attracted people suffering from rheumatism and arthritis until the end of the nineteenth
century. Around 1735, the spring became a mine; galleries were opened
in the hillside, and ten years later wells were dug to a depth of 80
meters to get at the bituminous asphalt, which was dug out with shov-

---

*The *sonde*, a technical term used in both French and English,
designates a device which carried electrodes into the borehole.
Wires connected the electrodes to a potentiometer on the surface, allowing measurements of the resistivities of the several
strata through which the borehole passed.

els. Not until 120 years later did Péchelbronn produce its first barrel of oil; real production began in 1882.

My father, whom I was badgering with questions, explained that what they had in mind was to take measurements at the bottom of a hole and record them in circumstances totally unlike those in the comfort of a physics laboratory. I could not follow him and wanted some sort of illustrative example, so he told me to imagine an eye, which, traveling up and down the length of a borehole, would recognize the nature of the different strata traversed by the bore. "Nothing could be easier to understand," he added. "You'll see when you join us at Péchelbronn."

Not far from Péchelbronn there was an establishment with some features of a hotel and some of a boarding house, where my father, my husband, and their crew made do as best they could. Monsieur Angel, the owner, was proud of the one bathtub in the house and was perfectly willing to have his boarders enjoy themselves in it, but as to putting a lot of hardware in it — woe to him who tried! Yet to a prospector a bathtub was more than a place to soak your feet, it was a tank for experiments. Dear Mr. Angel! Little did he suspect the black magic for which his tub was used.

Henri thought that with a good cable, a good winch, and good instruments, he would be able to get curves that would tell enough to confirm the exactness of the surface measures by those taken in depth. The Péchelbronn wells made it possible to achieve this vertical approach. The equipment, however, could not have been more rudimentary; roughly, it included three conductive cables spliced together with insulating tape every five meters, a sonde weighted with lead, a winch operated by hand, storage batteries, and a potentiometer. The combination of a delicate potentiometer and a big hulk of a winch made an odd marriage, but it was counted on to produce results. Alas! Several times, as the heavy sonde was being hauled up, the cable broke, so that after a number of attempts the trial had to be suspended. This first effort had taken fifteen hours and had allowed them to study the hole to a depth of only 140 meters. However, the well was 600 meters deep. It was useless to patch the cable — a stronger one was needed. After a lot of trial and error a new cable was put together, the hope being that it would stand every strain. A taxi rushed it to Péchelbronn. The atmosphere was feverish; it seemed that the future was hanging by this cable. So close to the goal, uncertainty reigned. Would the measurements taken in the field correspond to the theoretical calculations? Tense and trembling, my father — though he was then forty-eight years old — felt the anxiety of a student facing the test of his life. He was the youngest man on the team and the most eager at the job. He was as surprised as a neophyte by these experiments, which were about to make a giant step in the history of geophysics; he still found it extraordinary that so much could be learned with three bits of wire at the bottom of a hole.

A few days later, on September 6, 1927, in an inn in Woerth, a dinner marked a memorable date — the date of the first electrical measurements taken in a borehole. No longer was the purpose to explore the subsurface in order to discover ores hidden there, as had been the case with electrical prospecting on the surface. Now the sonde, lowered into a well at the end of a cable, had to measure and record, continuously and precisely, the physical properties of the strata traversed by the bore. Since the resistivity of porous rocks impregnated with oil or gas was greater than the resistivity of those filled with water, the "eye" mentioned by my father should be able to distinguish between the two kinds of strata. This was the birthday of "electrical coring,"* so called by analogy with the geological samples called "cores" which are taken by machine.

In Houston, where he now lived, Léonardon thought about his American job. He was convinced that in America lay the future of *Pros.* He settled in, took root. Gallois and Baron joined him. They rented an office, equipped it with furniture bought at sales — and for clothes . . . well, one could get along with two shirts. All that didn't cost much; Paris ought to understand. Paris groused . . . O.K., O.K., but watch the expenses. The fact is that returns were not enough, far from it — so far. Actually, there was some thought of getting one of the big American companies to finance electrical coring: $50,000 (let's say) and shares. That, however, would be selling the goose that was about to lay golden eggs, and for a pittance. It did not happen because there were no takers,

---

*The term *electrical coring* was adopted in 1927 by analogy with mechanical coring, which consisted of analyzing samples of earth (cores) cut out and brought up as drilling proceeded. The expression was replaced in 1933 by *electrical logging,* the term now in use throughout the world of petroleum. The word *log* designates a strip of paper or film on which measurements are recorded as curves (diagrams) in terms of depth, as a marine log measures the distance traveled by a ship in time. In conformity with chronological development, we shall employ the expression *electrical coring* for the process in use until 1933 and *logging* after that date.

or because these people, who spoke English with an accent, did not inspire confidence.

Between Houston and Paris, letters and telegrams passed each other in the mid-Atlantic. Things were not always measured with the same yardstick on both sides of the ocean. Léonardon champed at the bit. Wearing his habitual grey suit, loquacious beyond belief, he always managed to have the last word. One day, having attended a lecture by a certain Dr. Mason (a university professor who was considered a leading light on electrical prospecting), Léonardon convinced himself—if he needed convincing—that those *bons messieurs*, as he called Conrad and Marcel, knew infinitely more about the subject, and that the time had come for them to make themselves known by publishing scientific papers in specialized journals. Keep people in suspense, let them know that "we" exist; to talk about oneself was fair play. No question, of course, of giving information to competitors on the *qui vive*, but there was nothing to keep Conrad and Marcel from saying lots of exciting things laced with mathematical formulas without giving away the whole show. By omission, they lied, a little. Since Léonardon knew that these *bons messieurs* would bristle at the mere suggestion of false pretenses, he added with a touch of sly humor, "Obviously, it would cause you suffering if you had to sign such things, but that sacrifice is not required of you. . . ." Certain that he was on the right track, Léonardon confronted one with a *fait accompli*, however you might feel about it. And, if he kept you informed at all, it was strictly *pro forma*.

The only way to take America, according to Léonardon, was head-on and at the double. He proposed to rent an office in New York—no, he had already rented it, right on Broadway, the Cunard Building, in the heart of the financial district. Don't worry, not too high-priced, Messieurs, only $125 a month, and we'll manage to share secretaries with other offices on the same floor—that's what's good about Americans, they're ready to share costs. "Our modesty will kill us," he wrote, giving the wheel an extra push.

Paris, indeed, *was* conservative by nature and conviction. They were studying the basin of Briey in Lorraine; were sending Poldini to Spain, where there was a question of a geophysical study on the Guadalquivir fault; were getting good results in lignites in the Landes; but, first and foremost, electrical coring had to be perfected—in short, they had to improve the process rather than grow too fast. Time was needed, and perseverance . . . capital, too—hence the idea of selling or licensing their patents in order to get some money out of them.

What Conrad needed was to be sure of his lines of supply, to feel the ground under his feet. One would have thought that this man, who was literally consumed mind, body, and soul with the work at hand, was secretly — almost unconsciously — afraid of taking a false step. Marcel, being more aggressive, refused to put his signature to papers that might tie up the future.

In any case, Léonardon was informed in September 1927 that electrical coring had been tried out successfully. He was beside himself with impatience. He wanted the coring equipment right away; it was not in his nature to dillydally. America was his for the taking!

The first corings done in American soil were failures; neither the apparatus nor the engineers were quite ready for the job. This setback did not encourage the oil companies to go on. But even though this second attempt to conquer America was a fiasco like the first, Paris did not doubt that effort and perseverance would, as always, bring results. Faith in the fundamental correctness of their research and a strong tenacity were in the blood of these men. Conrad, Marcel, and Henri did not have it in them to give up. Looking back over the years, I see now that no one had foreseen — nor even suspected — the extraordinary future of electrical coring. They worked at it, of course, but as a kind of complement to surface measuring, to which the prime effort was always devoted.

In the United States, once the Roxana Petroleum contract was canceled, there were several crews looking for prospecting to do. As I followed them on my wall map, they seemed like the *compagnons du tour de France* — those wandering artisans of yore who went from town to town in France, tools over their shoulders, offering to work for anyone who would hire them. But our men were in a foreign country seventeen times as big as France, and their tools — an entire set of equipment shipped overseas at high cost — weighed too heavily in the balance to remain idle. Paris, already thinking of bringing these people home, had to parry a barrage of letters and telegrams: Léonardon refused to let go. He turned a deaf ear to the complaint that his stubbornness was draining the resources of the enterprise.

This little American show is your show all the same. I can understand that my reports look discouraging to anyone to whom victory bulletins alone are acceptable. But, mes bons Messieurs, do me the favor of not losing sight of the goal. One thing or the other—forget the whole thing or let someone who thinks he knows his business get on with it. . . .

Modest Léonardon! His America ... it isn't that he *thinks* he knows it; he knows it as if he had made it. He was a one-man band, playing twenty instruments at once. He bought the directory of mines, took down two thousand addresses, sent that many letters extolling electrical prospecting; he knocked on doors, caught drillers and geologists by the sleeve, jumped from one train to the other, broke in on his fellow travelers in the Pullman car — "Excuse me, would you by chance be in oil?" Nothing and nobody could put him off. He was the first to laugh at this habit of beating his own drum — the bass drum predominant. He got hold of Sherwin Kelly again — Kelly, who had a way with people but whose past success had given him such airs that Marc Schlumberger, talking about him, used the term "megalomaniac." Léonardon, a shrewd and energetic person himself, did not much take to Kelly, but since Kelly knew a great many people and made his way easily in mining circles, he hired him anyway, prepared to keep him on a leash. Then, to avoid standing still, Léonardon went to Canada, opened an office in Toronto, and deposited several crews in the farthest reaches of Ontario.

In these northern regions, where hundreds of miles often lay between the prospectors and the nearest hamlet, climate and terrain tested men to the limit. One encountered a succession of lakes, forests, swamplands, and places named Kamiskotia, Trimmins, Half Moon Lake, Hébécourt, Duparquet.... Although in late July they were beginning to disappear, clouds of gnats and midges ate you alive; the following winter, the people who had suffered those insects preferred the bite of the icy cold. Later, when working for the International Petroleum Company (at Turner Valley near Calgary, in southern Alberta), they would travel by sleigh. Now, however, under the rains of the waning summer, Ontario was nothing but trails and mud, mud and trails. In order to drag the cables in the forest they had to clear a path with hatchets, so dense was the vegetation. Nothing was easier than to get lost or to come nose to nose with a bear. After examining the outcrops of rock one evening, André Allégret and a Canadian geologist went round and round for several hours before they found their tent, which had been nearby all the time.

During that particular mission, Allégret and his crew spent three months in almost total isolation. Knapsacks on their backs, carrying equipment and provisions, the various crews started out, made their measurements, and went on. Sometimes their exploration brought them to horrible camps or to sordid "hotels." The prospectors, in order to compare the results of their work, met in one place or another after arduous marches, hours of canoeing broken by portages, and nights spent lying on the ground which was already beginning to freeze. Their one diversion was fishing, which was excellent. Returning to Toronto, they wrote their reports and described the experiences: " . . . morale is good in spite of living conditions, in which the work is our only link to the past. Prospecting makes us understand better what we were before — timorous men, held down by middle-class conventions and trivial reading. . . ." Yes, I too understood better, although I shared their existence only through hearsay; the winds of the great open spaces which whipped them, swept over the routines of my own life.

This first Canadian prospecting expedition had lasted only three months when Paris came charging in again. Compared to the results, it was all costing too much — in men, equipment, and scattered effort. Granted that Europe was not Golconda, at least it provided work that made more sense — paying less, it was true, but making survival possible. Once again, the old argument was before the house: Rather than trying to do several things at once, the company should give priority to improvement of techniques. The conquest of America would come as a bonus, in good time and at lower cost.

Léonardon refused to be terrorized so easily. Why were they talking to him about undue expenses when he was talking about winning a world? He quibbled a bit, clinging as though for dear life: "Let us hold on for a year," he pleaded, "hold on for six months, but hold on!" He could barely refrain from swearing. His crews, he protested, were hard at work. Their missions took them from Massachusetts to California, from Connecticut to Arizona — mines here, dams there, petroleum somewhere else. The proof, he cabled triumphantly in January 1929, was that his balance sheet showed a profit of $12,976.28. There it was, 28 cents included, not one cent less!

6

Blue Pins on a World Map

IN Paris, the company's offices and shops were growing larger. Given the speed with which *Pros* was developing, it was obvious that more space was needed. An apartment house and a small third-class hotel were bought and connected with 42, Rue Saint-Dominique. Because plans and drawings were my uncle's domain (he having both a predilection and genius for such things), my father, who was bored by the building process, left the remodeling entirely in his hands.

By 1929, the staff had grown from 58 to 95, of whom 53 were engineers. The profits, nonexistent until then, amounted to 14 million francs. The Depression in the United States was beginning to affect Europe, yet morale within *Pros* continued to be high. As I was trying to follow the movements of the crews in an atlas, my father had a map of the earth hung on the wall for my use; I studded it with pins that had blue porcelain heads (blue being my favorite color). I was playing a tactical or strategic game—I'm not sure which—following the warlike bulletins I insisted on receiving:

Soumont (Normandy)	*Finished. Went well.*
	Will start again in September . . .
Landes	*Work progressing. Potassium.*
	Two teams. Results uncertain.
Benisaf (Algeria)	*Finished. Drilling to 50 m.*
	Several lodes of ore probable.
	400,000 francs premium almost assured.
Spain	*Finished. Half success.*
	Ought to be reopened.

Péchelbronn	*Success in sight.*
Katanga	*Three crews at work.*

And so on. But — games notwithstanding — I knew that I was dealing with men of flesh-and-blood, men whose lives I hoped to follow to the four corners of the world.

For lack of time, the training of engineers was on the summary side. After some weeks at Péchelbronn, learning to use the apparatus and the potentiometer, they were considered ready to venture out on their own. Armed with their equipment, a small travel notebook, and a large dose of good advice, they took to the road like so many missionaries to preach electrical prospecting. They were advised not to adopt the spendthrift habits of the Americans — a source of immediate but extremely relative satisfaction. On the contrary, they were expected to combine Protestant frugality with Yankee know-how, and to give the best of themselves in order to get the job done. As I listened, it all seemed a weird moral equation to me: An engineer represented a small amount of capital to be put to work and the capital, even a potential sum, had to produce.

Who were these young men whose fortunes touched me to the quick? What heads, blond or dark, were hidden under my porcelain pinheads? From what past, what milieu did they come? By what criteria were they chosen? "Oh," I was told, "they come from everywhere, and if they look good and have a decent degree. . . ." Later on — in 1932, I think — an advertisement placed in the *Revue des Elèves* of the Ecole des Mines in Paris amused me:

Throughout the whole world,
Under the sea, on the moon,
Proselec discovers
The structure of mines
To offer you a mine . . . of jobs

I never found out what Sunday poet had authored this gem, which smacked of mystery. The fact is that most of the personnel were recruited at the Ecole des Mines, Polytechnique, Centrale, the Ecole Supérieure d'Electricité, and Arts et Métiers. The majority came from the middle class. Other than "good looks" and an acceptable diploma, what these budding prospectors had to have in common was a taste for strange places and life in the open air, and — most important — a desire to be engaged in work not organized on traditional lines, but work in which individual initiative and resourcefulness marked the road to success.

After the training period, chance often determined who would go to one country rather than another. I remember that in 1929 Pierre Bayle and Raymond Sauvage found it difficult to choose between Venezuela and the Soviet Union: Stalin's country aroused curiosity, Bolivar's was not unattractive. So they tossed a coin and Venezuela fell to Bayle, the U.S.S.R. to Sauvage.

So keen was my desire to learn, so far-ranging my imagination, that the map on the wall soon left me hungry for more. I began to ask questions of the prospectors who passed through; I read the mail that came in from faraway places, pausing to study the seals and postage stamps, the mere look of which somehow made me feel that I was "there." It was just that: I wanted to be with those young unknowns who did not know that I even existed. If a letter arrived from Canada or South Africa or California, I quickly surrounded myself with books about Canada, South Africa, California. The history of those lands and their flora and fauna took me out of the routine of office work. Every morning, the minute I reached the headquarters on Rue Saint-Dominique, I went straight to "Maté" Provost, the secretary (whom I never saw, it seems to me, without a letter opener in her hand), and watched her open envelope after envelope. Roly-poly, well turned out, calling each person by name, she proceeded to distribute the mail. It was a ritual: She held a kind of literary salon, where the works under discussion were signed Lalande, Poirault, Piotte.... But before each one received his or her share (and I my manna), my husband gleaned the tales of electrodes, cables, and depths of investigation that made up his life.

My sister Dominique, with her degrees in mathematics and physics, felt toward me the rightful superiority of the graduate student. The editorship of *Proselec*, the confidential in-house magazine that assured technical communication between the different crews, was left to her; dreams and escapism were mine. When I heard her talk of logarithmic tables and overburden, it was all Greek to me; when I talked about prospectors lost in the forests of Canada, it was all verbiage to her: So we were even.

I wonder today how much of my reading or my interpretation of the stories I heard was colored by the romanticism of a young woman scarcely out of school. Bulgarian forest ... Hungarian *puszta* ... Russian steppe ... the pull of the faraway, of the other side of things: Out there, wherever "out there" was, the water and the bread had a taste that I almost begrudged our travelers. I had not the slightest doubt that the winter might be rainy and the houses glacial in Andalusia; but the rain was not the same as here, nor was the cold, nor were the houses. Out there they warmed themselves around a terra-cotta pot filled with live coals. They put candles on the measuring stakes to get their bearings at night. A gypsy assistant dragged the cables in the olive groves and blew on a conch shell like a shepherd of old. The engineer heard the note and closed the circuit. The potentiometer needle came to rest as silence reigned again. . . . Charles Scheibli, who had the good luck to be there and whose letters I read, wrote: "Monsieur Conrad told me that 'well-off folks' paid high prices to spend the winter in Andalusia. I tell you frankly that I would pay to get out of it." But this Charles Scheibli was a great humorist.

No! there were no stormy skies and no untoward occurrences in the world of my imagination. My father was thrown by his horse while he was spreading out a map on the animal's withers? Well, so what? No harm had come of it, since it happened in Bulgaria! A mule laden with prospecting equipment fell over a cliff? Well, it's a sure thing that he landed in a bed of moss, since it happened in the Pyrenees! An engineer got himself jailed because his passport was not in order? Ah, but his cell must have been comfortable and cheery, since it happened in Venezuela! Perhaps. But what matters most is that I was protecting them from harm by dint of nothing other than my imagination.

As subjective as my mythology may have been, it was nevertheless anchored in reality. Father, husband, uncle, the men at Rue Saint-Dominique and the men out in the brush, occupied a real space in which my imagination took root. After all, everything I invented for myself was fed from real sources. The first source was the Rue Saint-Dominique office, that conglomeration of floors on different levels, a made-over bistro, hallways, cellars with a well and a tank for experiments—altogether a higgledy-piggledy architecture. Seeing this for the very first time, a young engineer coming to work would probably think it unlikely that anything worthwhile could be accomplished in these surroundings. It was indeed a funny sort of shop, where nobody bore an official title, where a kind of Brook Farm good fellowship was the rule, where engineers and workers chatted as equals, where the "veterans" were legends in their own lifetimes, where everyone—man or woman—seemed to come out of a mold so unconventional that at first sight it was hard to know which way to turn. Everything went on as if there were no one, from doorman to boss, who was not a person in his own right.

Who was there? Well, for instance, there was Jules Miller, who was in charge of finances and kept the accounts . . . a man who while still young looked old, and who, according to Boris Schneersohn (who worked beside him for thirty-five years), looked quite young in his old age. Jules's voice sounded angry, his manner dry and brusque, and he was all the more ready to let you into his office because he knew it was beyond him to refuse you a service. There was René Clairin, who headed the research office and worked closely with Marcel. Never were two men (superficially, at least) less alike — the one distinguished by an aristocratic ease of manner, the other hardly noticeable. Yet no one else could catch one of Marcel's ideas on the fly as Clairin did, or more quickly draw up a blueprint for it, or put it into execution more rapidly. Clumsy of body and awkward of speech as he may have appeared, Clairin was perhaps the only person who took the liberty of differing with his chief.

There was Paul Charrin, who was in charge of the crews in the field, who seemed to grow taller and thinner as you watched, and whose fits of wrath shook the very walls; and there was Raoul de Geffrier, his assistant, as imperturbable as the other was easily aroused. There was Henri Doll, who handed you an invention every time you turned around. "Between Charrin and me," Henri told me, "we could keep the shop running."

There was Georges Lahaye, the only mechanic in the testing department (all the others were engineers), who, reigning there as undisputed master, meted out a fine of a round of drinks to anyone who broke the rules of his fief. And there was Bourumeau, the concierge, dapper and trim, who also ruled his particular domain with the airs of the boss man himself. So, in 1934 or thereabouts, there were sixty or seventy people, of whom the greater number — experienced prospectors and young beginners alike — were going through a period of training or retraining before setting out on their various missions.

This constant movement created a climate in which cheerfulness in no way impeded the stirring up of new ideas. I marveled at the way these people, each with his or her strongly marked personality, could live and work together without friction and do their thinking without interference. I know that my private dreams must have colored my impressions, but there really was an atmosphere of work and celebration which was felt by others than myself. The newcomers, the non-initiates — I almost said the laymen — had some reason to wonder whether this *Pros* (no one quite knew what it produced) was not a friendly association of dilettantes rather than an enterprise deserving respect. They found here a kind of overflow of freedom and a heady enthusiasm — so much so that if Tom, Dick, or Harry had an idea, everybody acclaimed him as a genius. What profusions of praise came when someone hit on a promising twist! At once they examined it inside and out, built it, and tested it — and then what fireworks if the new discovery belied its promise!

I have long wondered how this climate, in which hard work and high spirits went together, came about. Every new recruit — worker or engineer, stenographer or stockkeeper, even the crank and the scowler — was caught up by this spirit. Everything went on as if work and drudgery were not synonymous. Of course, the time had passed when the Ecole des Mines was kind enough to allow my father the use of a cellar where he could conduct experiments deemed of little value. But was I the only one to remember that past, recent as it was? Was I alone in hearing its echo, now that the bathtub was in a fair way to take on the dimensions of the ocean, and the piles of sand and clay those of the continents? This undertaking which was still seeking its way, inventing itself with the world for its laboratory — whence came its character, so unconventional that some thought it mere adventure? Certainly there was no question of a graft, of some input from outside; it was Pros's mark, its natural bent, as when an acquired trait little by little becomes a characteristic. In a factory or office, team spirit dwindles as the work is fragmented and the task of each worker becomes more limited; the pleasure of contributing to a common task is seldom found there. But our equipment was so uniquely conceived and built that, on entering this ill-defined market, it did not lose its individuality. The tie between the equipment and the man who made it was never cut.

Because everyone contributed on a footing of equality, and because the discussion of ideas, methods, and techniques was given free rein without the constraints of a hierarchy to stifle spontaneity, a rare cohesion welded "thinkers" and craftsmen, "administrators" and prospectors, into a solid unity. The prospectors, coming back from Alsace or the ends of the earth, could go straight in to the "bosses," tell them about their work and experiences, voice their criticisms, make their personal problems known. They communicated to the engineers, technicians, and secretaries a feeling of the wide-open spaces — of adventure. What added to that cohesion, I think, was the low ratio of workmen to engineers — four or five to one — whereas in industry it was from fifty to two hundred to one. And if, as was true, my father and my uncle quite naturally practiced an "enlightened" paternalism, it was just as true that *Pros* was a shop unlike others.

It was different, in fact, even in the day-to-day work. Everything went along as if fun and laughter were part of even the least likely tasks. One day, when the core bullet was ready for use, those who had worked on it wanted to test this "sample taker" for speed and trajectory; they decided that firing it into the pool in the courtyard would do the trick. The experiment was splendid and spectacular. But the second firing drove everyone at *Pros* out of the building and into the courtyard. The water in the pool being uncompressible, the whole building had been shaken by the detonation; even the bust of Napoleon in the corner bistro was split down the middle. Yet the tests had to proceed, and my uncle suggested that there was no better way to make the water elastic than to weigh down some toy balloons — inflated, of course — and place them at the bottom of the pool: The detonation would burst them and the shock would be deadened. In two days there was not a single balloon left in the Champ-de-Mars *quartier*; one week later there were none in all of Paris. Never mind! What children's balloons could do, housekeepers' pails could do just as well: Weight them two by two, lower them upside down to the bottom, and they would carry down just enough air to act as deadeners. But there again, those fine utensils could not stand up under the treatment they were subjected to, and in no time the shopkeepers in the *quartier* were out of them. Imagine the hurrahs when somebody returned after a fruitful hunt, showing off a pair of pails like a glorious trophy!

This ability to manage with whatever came to hand was, so to speak, the trademark of *Pros*. It seems to me traceable to the time when I, as a little girl, watched my father reproduce the geological strata of the earth's crust in a child's bathtub. A procedure was born there and also an unwritten tradition, like the customs that so infiltrate a way of living that they become second nature. And it is true that the sort of never-failing ingenuity that abounded there was handed on from one to the other, like the laughter that accompanied it — the laughter that tempered anxiety about taking the wrong road.

7

Westward Ho!

AMERICA ... land of milk and honey, *Pros*'s future, Léonardon's game preserve — I had heard so many stories about it that I could not wait to land. On July 3, 1929, after ten days aboard the *De Grasse*, New York appeared out of a heat haze. The Statue of Liberty with her torch, skyscrapers looming up to meet the ship, tugs with their throaty voices, passports, rubber stamps, luggage, porters; Léonardon with his ample gestures, unconcealed impatience, and torrential speech was on the dock.

The sight of my husband filled Léonardon with joy. For a long time he had been pleading for a visit from Henri Doll, the only man in the world (with the exception of *ces messieurs*) in whom he was willing to recognize the virtues required to replace him at the top of his American fief long enough to let him take a well-earned rest. But care and foresight never hurt anyone; so, in the cab that went plunging through the straight streets, he poured out an inexhaustible stream of good advice. As for me, I could not see enough — the docks, the business section, the neo-Grecian temple of the Public Library, the crowds in shirt-sleeves. ... It seemed that we were hurrying past a continuous stage setting. Tudor City, at the end of 42nd Street, the East River, and a patch of grass ... I was enjoying all of it already, but the price of the studio reserved for us on the eighth floor included only two beds in one closet, a kitchenette in another. It cost too much to rent even a patch of sky. I sank into a chair and wept.

At 25 Broadway, on the twentieth floor of the Cunard Building, a brass plaque designated the offices of Schlumberger Electrical Prospecting Methods. Two small rooms, work tables, a few odd chairs, a typewriter, a telephone covered with dust — but the oil companies would soon be calling. "This just proves that we exist, and that's something," said Henri. "Our methods and our work will do the rest."

41

Never had sweeter music come to Léonardon's ears. To him, Henri Doll's arrival promised the early establishment, on American soil, of electrical coring — a discovery (he was sure of this) more important than the invention of the wheel. A letter from Marcel tried his patience to the breaking point. If electrical coring took hold in the United States, Marcel was thinking, why not get a drilling company interested in the rights — let's say $50,000? Fifty thousand dollars would be very welcome; the treasury would be back in the black.

Léonardon had a fit. "Absolutely not!" he thundered. "Give up the orange for the rind?" The minute he was back in France he would show what they had to offer, and with my husband behind him ... well, he'd bring down a shower of dollars. Three months — that was all he needed to win the bet. The secretary's lacquered fingernails on the old Underwood typewriter made a racket like hail falling on a tin roof.

The day after we landed, Henri showed me around New York. It was pleasantly cool as we rode on the top of the double-decker buses. Who has not seen, in picture books or in the movies, the canyons of downtown Manhattan; the lacework of the Brooklyn Bridge; the New Jersey Palisades rising steep on the west shore of the Hudson River, that great estuary whose fresh water mingles with the salt water of the Atlantic? Level streets, vertical stone, barbaric architecture, commercial utilitarianism — all that is pure cliché. To me, the city looked beautiful, strangely abstract. Also abstract was the office on the twentieth floor, where I felt out of my element and sought relief in studying the wall maps studded here and there with blue pins. These at least brought me back to the concrete existence of the prospecting crews. Texas, California, Canada — the morning mail brought its daily ration of electrical measurements, which my husband (in order to make things clear to the companies) transcribed into linear language. But how to expound the secrets of your science to people whose speech is not your own? Henri's English, which would become flawless in time, was then rather poor. So he had to fall back on Sherwin Kelly, the man of a thousand connections, whose nimbleness of tongue was such that he could have peddled electrical prospecting like the pushcart peddlers whom I saw hawking their pretzels at the street corners.

To Henri, nothing could have been more foreign than this hard-sell approach; the only thing that interested him was technology. It is hard to imagine a mind so detached from (I almost said forgetful of) any field other than its own. That Conrad and Marcel, if pushed, could have given a different direction to their lives — this I find conceivable. Not Henri. Never had a human being been so right about his vocation.

Maybe he did have only one string to his bow, but he knew how to play it to perfection. He planned his shot from afar and in depth; when he let fly, his arrows hit the target like a hailstorm. Since he chose his target knowing precisely what he was doing and had calculated his angles of approach with meticulous care, his case was already won — or at any rate well advanced — when he formulated his guidelines. No one could have pleaded his own cause better or with more spirit, and this, as one might expect, at times led to some friction.

Not that Conrad or Marcel wanted to "boss" Henri: His theoretical and practical contributions to prospecting methods had quickly made him their favorite collaborator. Yet the fact remains that they ran things from a certain height, Conrad devoting himself to his own ideas and Marcel to practical developments, with Henri between the two — a mediator in spite of himself — in search of suggestions that would put all three on the same wavelength. During the second half of the 'twenties, when the distribution of tasks was influenced by the scarcity of resources, Henri saw his energy being drained off in directions that ran counter to his instincts. His duties were not only the shop, the field-work, and the research, but also interviewing and at times instructing newly recruited engineers, dealing with suppliers, and keeping accounts of equipment and production costs (to which end he had taken courses at the Ecole des Sciences Politiques). So that it was this trip to the United States — insistently requested by Léonardon and long resisted by Conrad and Marcel — which gave Henri a renewed sense of being his own man.

Every morning we were swallowed up by the subway, one of us going downtown, the other uptown — Henri to his curves and logarithms, I to my English classes at Columbia University.

Columbia . . . its imposing edifices with their Doric colonnades, standing around lawns yellowed by the torrid sun of July, its monumental flights of steps, its notables draped in their bronze togas, its inscriptions above the porticos exalting the light of reason and the benefactions of science. I was not looking for all that. What I wanted was to learn the language and to meet Americans my own age. I was disappointed to find that the summer courses attracted few students with American roots. True, I met one young woman from the Midwest who was working for a degree in journalism; but I also became acquainted with a Greek priest, several Italians, a Polish couple, a Chinese, some Spaniards — all immigrants on the way to becoming American citizens, all talking to me nostalgically about islands in the Aegean Sea, or Apulia and Tuscany, or flamenco, or Cantonese cuisine. All the foresight in the world could not have warned me that I myself would endure this homesickness a decade later.

In any case, I found my life as a student agreeable: having a sandwich for lunch, sipping a brownish beverage ("American coffee") from a cup that felt like a beer mug, reading whenever and whatever I liked, filling my time by observing tiny details. About five in the afternoon the subway took me back to our apartment, where my husband joined me soon after. We shared our day's experiences, had a snack, went out and hit balls around a miniature golf course; our imaginations were as limited as the budget. Sunday was our "day off"; we would go for a long walk in Central Park, or sail around Manhattan on the Circle Line aboard which the music of Central European accordions and the aroma of sausage rose in the almost-cool breeze. This folklore of the poor — the only folklore I came to know in New York — made an unforgettable impression on me.

Several months passed in this manner. I had still seen nothing of America except New York and time was beginning to hang heavy on my hands. Then, one day, when autumn was already decking New England with the loveliest shades of copper, red, and gold, my husband drew a plan for a real journey: We would go to the oil fields of Oklahoma. At once — though to me that area was no more than a legend — I was consumed with impatience. My English had improved — not much perhaps, but enough to make possible a lot of haphazard reading in the Columbia library. I wanted to learn and remember everything about the states we were to travel through for two days and two nights on the train.

Alas! Precisely because of my insatiable curiosity I absorbed little more than scraps and snippets of knowledge. New Jersey, about which I remembered only its nickname, the "Garden State," dismayed me by its lack of charm. What I saw from the train was swampy flatlands, punctuated here and there by industrial complexes and urban centers, all alike and dismal under a steady rain; the gardens, if there were any, must have been over the horizon. It was not until we got to Pennsylvania, and especially after we were west of Philadelphia, that the vegetation came into its own — the land arable, the hills home to flocks. I loved the white houses and red barns on the widely separated farms. I wondered whether we were passing through the Amish country — the land of Mennonite farmers with their thick beards, black clothes, and horse-drawn buggies, to whom the black smoke belching from our locomotive was an emanation of the devil. Much farther to the west, after nightfall, the steel furnaces of Pittsburgh set the sky on fire. I knew that the first oil well had been dug in 1859 about a hundred miles to the north, at a place called Titusville. The yield (eight or ten barrels a day) was hardly worth the trouble; but two years later, about a mile away at Oil Creek, a huge spout of oil burst forth at the drillers' feet, caught fire, and killed nineteen people. That had happened sixty-eight years before, yet the glow over Pittsburgh seemed to reflect the glare of the earlier conflagration.

After leaving the Allegheny mountains behind us, we traveled over the rolling plains of Ohio, which not long ago had been covered by vast chestnut forests. An interminable halt in the station at Columbus; another at Indianapolis, after a two-hundred-mile run along an unbroken plain—one of the most fertile anywhere. With my nose glued to the car window I fought off sleep, not wanting to miss anything. I was sorry the train did not make a detour to the North, where Lincoln once lived. Two hundred fifty miles or so farther on, we turned southwest through central Illinois and crossed the Mississippi at St. Louis. Now we were downstream from the Missouri River, Mark Twain country. Farmland as far as the eye could see, lakes, streams, the Ozark plateau (sparsely populated by people whose way of life belonged to another century), and then, still to the south and west, four hundred miles away, Oklahoma at last—and Tulsa, on the banks of the Arkansas, where Gilbert Deschâtre was waiting for us. We hardly had time to shake hands before the two men were launched into a technical discussion.

We stayed in Tulsa for one week. Crates of equipment shipped from Paris had arrived there before us. They contained an electrically driven winch and an improved recorder, both devised by Marcel. While I almost died of boredom in this not particularly attractive city, my husband and Deschâtre were busy mounting the winch on the chassis of an American truck (which was obviously built for it) and equipped it with a power takeoff that would make it turn.

The end of November was at hand. The weather was dry and cold the morning we left for Seminole. While Deschâtre, who was driving, went on and on with his shop talk, the jolts and bounces of the truck made me think of the stagecoaches that had used these deeply-rutted roads not long ago. Twenty-two years earlier—in 1907—this vast plain, the northern part of which stretches without a tree in sight all the way to the foothills of the Rockies, was still Indian Territory. Cherokees and Choctaws were still at home in Oklahoma: the state name means "Land of the Red Men."

The hamlets we passed through had names like Okmulgee, Okemah, Waleetka, Wewoka. Seminole, with its muddy lanes bordered by nondescript buildings, had taken its name from an Indian people native to Georgia and Alabama; some of them had settled in Florida, some in Oklahoma, following the Seminole War (1835–1842). In 1929, black gold uprooted the Seminoles once again—and far more insidiously than in the past. The boom was so great and drew such crowds (oh, high-spirited hopefuls) that finding a place to lodge was a chancy business.

We arrived in Seminole after nightfall and walked into a saloon that had about it something of a gambling den, something of a fancy house. Our entrance caused a great stir: The "ladies" (to use Henri's euphemism) who usually ventured into such places were not exactly the type who accompanied their spouses to take measurements in bore-holes. An atmosphere thick with tobacco smoke and alcohol, loud with thunderous voices and the tinkling of a piano player, enveloped me; then, before I could get my bearings, I felt myself being hoisted from the floor and deposited on a table like a Chinese doll. Tearing them-selves away from their fanciful schemes (which would make million-aires of them in the time it took to dig the right hole in the right place), several strapping fellows devoured my skirt with their eyes, as children would a birthday cake. I was the *chanteuse* or the dancer they were waiting for, of whom they expected great entertainment.

Not I, alas! I was only the wife of a graduate engineer on a field mission. Nothing funny about that, nothing funny at all for lonesome oil drillers. My husband was laughing at their frustration and found my embarrassment amusing. As for Deschâtre, the poor man, I remem-ber a quick surge of annoyance at seeing him there, falling asleep on his feet as if nothing at all was happening. However, when I was leav-ing New York I had read his latest report to date. For the period from September 15 to October 17 alone, he had to his credit 1,760 kilometers by truck; 21 kilometers of cored wells; 13,600 measurements; 2,600 meters of cable lost in the boreholes; not to mention talks with geolo-gists from the oil companies, maintenance and repair of equipment, 10 reports, 53 telegrams sent or received—in short, an average of 15 hours of work per day. "Want a job!" he had added, and what a month! It did not take me long to see that he was always short of sleep; nobody could match his way of seizing the smallest opportunity to burrow into some unlikely hole, where only his loud breathing gave his presence away.

The workmen—big, husky men in boots crusted with greasy mud—lodged in rickety shanties built of boards hastily slapped together. Their only diversions were drink and the company of frowsy prostitutes. To me the workers all looked as if they came from the same

mold, so that I could not tell them apart except by their motley clothing. There were always some who accosted me in passing, and the bold-
est of them (or maybe the most desperate) stood watch at the door of
my room. Yet they were clumsy, awkward types, more timid than
intimidating: After all, was I not a "lady"? On the other hand, the
streetwalkers did not stand on ceremony; they barely stopped short of
sending me word that I was giving them unfair competition. I certainly
found it harder to reassure them than to discourage my suitors. As for
the Indians, they never even turned their heads my way. Anyway,
there was nothing "Indian" about most of them except their square-cut
features and their quasi-religious silence. The white men had brought
them firewater and syphilis and reduced them to the status of helots.
They had seen their feathered headdresses, their bows and arrows, their
tomahawks and wigwams become exhibits in anthropology museums.
The few who had succeeded in riding the crest of the wave looked like
extras brought on to lend "authenticity" to a Western film. Here and
there I saw men who, smoking dollar-apiece cigars as if they were peace
pipes, looked like nothing at all, not even like musical-comedy red-
skins. They were the ones that petroleum had enriched and corrupted.
Begrimed with oil and wearing green jumpers, they could hardly be
distinguished from the derricks that lit up the night with their diabol-
ical flames. To me the derricks seemed like demons, working the metal
shafts that went up and down, clashing against each other in a hellish
symphony.

My husband and Deschâtre toiled over their apparatus, which did
not always give out the hoped-for measurements that would reveal
deposit of hydrocarbons. Until then, before the invention of electrical
coring, the only method available for the study of the formations traversed by a drillhole was to take out cores here and there with a core
barrel; the cores were then subjected to chemical analysis. The process
was laborious and not very accurate. The rock was hard and not easily
penetrated; it was also friable and crumbled under the impact of the
coring tool. The holes, necessarily spaced at some distance from each
other, did not give a true picture of the subsurface, and there was a real
risk of skirting a pay load without "seeing" it. In addition, drilling had
to be stopped while the samples were sent to laboratories—often very
distant—and the results of analysis came back; so the method could not
help being costly. Sometimes, in their impatience to find out what they
were getting at, the drillers and geologists licked the cores to see
whether they tasted of petroleum.

The fact is that the *Pros* engineers were the first in the world to succeed in making electrical measurements along the sides of a bore hole. Horizontal on the surface before, prospecting now became vertical. While the older method covered large areas of terrain, the new one limited itself to the strata in the immediate vicinity of the boreholes. Electrical coring did away with the delays connected with waiting for laboratory analyses; but more than that, the process made possible a continuous "seeing" of the different formations it examined. These were the measurements that my husband had come to the Seminole fields to take, using the equipment sent from Paris.

The first electrical coring in the United States had taken place on August 17, 1929, north of Santa Barbara, California, at a depth of 2,727 feet. Deschâtre, Roche, and Gallois spent several disappointing weeks there. Due both to the nature of the terrain and to the equipment in use, the results were far from brilliant. Weights were lost in the boreholes, the cable wore out quickly on the hand-driven winches, the tracing of the curves (also done by hand) left much to be desired.

Now, in Oklahoma, where the geological formations showed high resistivity and the borings were not nearly so deep, my husband hoped to "hit the jackpot" with the new tools available to him. Early in the morning we climbed aboard the truck. On other occasions our expeditions took place at night, in piercing cold. The roads, deeply rutted and covered with clayey mud by the continuous traffic of heavy equipment, were as slippery as a skating rink; more than once we went into heart-stopping skids, not daring to touch the brakes. The slightest upgrade or downgrade called for gymnastic maneuvers; and as for the wooden bridges that straddled creeks and ditches, they simply threatened us with catastrophe.

We were taking measurements for the Gypsy Company, a subsidiary of Gulf, but it happened that an "independent" wanted to take advantage of the opportunity to test his wells. One day, when my husband and Deschâtre were putting the final touches on their preparation for a series of measurements, a ferocious-looking, frenetic man rushed up to them, demanding to know what they were doing on his land.

Our men tried to tell him: "Well, M. Whatzizname, vice-president of the company that happens to be working this piece of land, has asked us to try electrical coring, and that's what we're . . ."

"To try *what?*" roared the stranger, who had never heard of anything but mechanical coring. "Try *what?* Look, I'm the president of that company! I give you five minutes to get the hell out of here, or else I'll lick the lot of you myself!"

As Henri—to whom euphemism came naturally—put it, "One

could see that he was a rather primitive man." We took off without further ado. It should be said, though, that tie drillers were always afraid that they were being spied on or that their wells would be sabotaged.

Each day brought its adventure, little or big. Once, we out of gas in the middle of the night and had to pick the lock on a closed gas pump—taking care to leave a note of apology and a ten-dollar bill, which must have amazed the owner. Another time, we broke down in the middle of a no-man's-land in twenty-below-zero-weather. And there were comical incidents, too, like the time the Maillot Hotel in Tulsa turned us away because we had driven up in a *truck*. Ironically, in this same hotel, years later, Henri Doll received the gold medal of the American Institute of Mining Metallurgical and Petroleum Engineers. But that hour of glory was in the distant future. For the moment, we had to fall back on a less fussy hotel, and this turned out well. The proprietor, a native of Marseilles, lavished attention on us in proportion to the pleasure it gave him to "talk French the way we did at home." He was married to a ravishingly beautiful Indian woman named Jaluca. The bright colours of her attire contrasted with my drab, poorly cut clothes. Jaluca was never without her crocodile-skin handbag, from which she extracted fat rolls of bills, laughing uproariously all the while. And I, incapable as I was of spending a dollar without feeling remorse and noting the expenditure in my notebook of accounts—as though to obtain absolution for an unwarranted waste of money—envied this charming creature's carefree ways and display of wealth. It may have been my brief contact with Jaluca that made me begin to rid myself of an ancestral inheritance that measured spiritual strength with the yardstick of parsimony.

The most unexpected adventure of all came from the chance observation of a phenomenon that appeared to be abnormal. One day, when my husband was working on some measurements near Seminole, he noticed an almost imperceptible trembling of the potentiometer needle when no current was being sent between the electrodes. When a second tracing of the curve taken at the same depth confirmed the observation, he jotted a memo on the back of the log: "Vibrations of the needles. S.P. probable." Six months later, in Paris, he came across this note by pure chance; it gave rise to an exceedingly fruitful train of thought. The day he talked to Conrad about it, Conrad told me in confidence that Henri had just put his finger on a discovery that might be decisive for the future of *Pros*. First laboratory research, then on-site studies (in Péchelbronn in 1930–31, Maracaibo, Grozny, and Baku in

1932), showed that *S.P.* (curve of spontaneous potential appearing without artificial stimulation) brought to light a phenomenon of electrofiltration due not only — as had been thought — to the difference in pressure between the column of mud in the borehole and the surrounding strata, but also to the differing degrees of salinity in the mud and in the water present in capillary form in the permeable strata. The resulting battery effect sets up currents of the electrochemical type which includes variations of potential. As Henri explained it to me, the curve of spontaneous polarization gives information on the permeability of the strata and the curve of resistivity on the amount of water or oil in them. It was to the remarkable complementary relationship of these curves that electrical coring owed its rapid success.

What comes most readily to mind about a short trip to Sunrise, Wyoming, some 800 miles northwest of Tulsa, is my memory of the polar cold. On the return journey, we stopped in Denver. There, while my husband talked about prospecting with the local geologists, I took refuge in the churches. I was surprised that each of twenty sects had its own temple in which to worship the same God.

Christmas was drawing closer. I was impatient to get home to my children; our travels had begun to pall. As Henri was returning to Tulsa, I decided to part company with him. Pent-up energy and a taste for the unknown also whetted my desire to take to the road alone. We agreed to meet again in New York in time for our departure by ship for France. Just as I had done on the trip out, I traveled with my nose pressed against the train car window. Arkansas went past with its rice fields, cotton plantations, and endless woodlands. At Memphis, on the east bank of the Mississippi, I got off the train. Here I was in the deep South, in Dixieland, home of the blues and Negro spirituals. What was I looking for in these parts? What idle taste for the exotic had brought me here? . . . I hailed a taxi and gave the name of a hotel to the driver.

Like St. Louis, but almost 300 miles to the south, Memphis anchored its docks in the turgid waters of the Mississippi. Stern-paddle boats churned up and down the river. The deckhands were black. Other blacks toted metal-strapped bales of cotton on their backs. I thought to myself that there was enough calico in those vegetable fibers to clothe the world, but not enough to cover the backs of these ragged men. Was it this same American cotton that my Alsatian forebears had woven on their Jacquard looms? . . . I had seen enough of Memphis and boarded the train again.

At that time, before commercial aviation, American trains were superlatively comfortable. Superlative also — to the point of being a nuisance — was the kindliness of my fellow travelers. When they found out (once the customary word or two was exchanged) that I was traveling for my own pleasure, that my husband was "in oil," and that I was from the land of Lafayette (pronounced "Laffyette"), the men could not be attentive enough nor the women sweet enough in their efforts to make me feel welcome on board. Exchanges of smiles, of stereotypical conversation, even of home addresses — how readily Southerners extend invitations!

The railroad branched off as we came into North Carolina and rolled along the foothills of the Great Smokies, their milky blue a color I have never seen elsewhere. Then came Virginia and its culture, more familiar to me because it was older — Virginia that had given four presidents to the nation, its tobacco fields stretching as far as the eye could see, its famous sites: Appomattox; Richmond, where the future of the Union was settled; the banks of the Potomac; Mount Vernon; and, finally, Washington, D.C.

Through the medium of a portrait, George Washington had long had a place in my mental landscape. From Ali Pasha to King Louis Philippe, from Metternich to Lord Aberdeen — I would have trouble enumerating all the personages who adorned Val-Richer, my childhood home. The American Congress had given François Guizot the portrait of Washington (a copy of one by Gilbert Stuart) and one of Alexander Hamilton, in thanks for Guizot's biographies of the two historic figures. At Val-Richer, when the time came for Sunday worship, the portrait of Guizot's mother holding her Bible made me look away toward the friendlier picture of Washington that hung on the drawing-room wall. I liked his vaguely conspiratorial smile: It was as if he was exhorting me to bear my boredom patiently.

My grandfather read the Bible to us, my grandmother read a sermon. Surrounded by her seventeen grandchildren my grandmother, Marguerite de Witt-Schlumberger, presided over their frolics. Presiding suited her perfectly. In fact, first and foremost, she was an indomitable suffragette. She had adopted the adage "What woman wants, God wills." The chairman of the Senate committee on suffrage once took it upon himself to tell her, "Women have enough to do taking care of their wardrobe." Her answer was, "Well, my dear sir, what women do you know?" Her activity on behalf of her cause placed her in the vanguard of the movement. She was elected head of the International Woman Suffrage Alliance, and her contacts with foreign feminist associations were numerous and continuous. One day — I must have been nine years old at the time — my grandmother had introduced me as a future suffragette to one of those ladies; the lady had exclaimed "Oh, you must come and visit me in Washington!" I still remember my speechless amazement upon learning that the general who smiled at me from his picture had given his name to the capital of the United States.

Anyway, here I was in Washington late in 1929, as the guest of a Democratic Women's Club, preceded by the fame of my grandmother and of my grandfather Guizot, who had been Prime Minister under Louis Philippe. My visit under these auspices was due to the initiative of Mrs. Carrie Chapman Catt; for it was she who, fifteen years earlier, had invited me to visit her in her home city. Today I cannot recall why this invitation charmed me, nor exactly what I expected of it. I suppose it revived the already dim image of my grandmother and, indirectly, certain childhood memories.

Mrs. Chapman Catt was too old to show me around in person, so she turned me over to the tender care of her secretary with a program so tightly scheduled that it ruined my enjoyment of being in the national capital. I have no doubt that these clubwomen were in turn disappointed by my indifference to their long-winded discussions. Too bad, but I did not have the bent for politics that my family background had led them to expect. I would have liked to discover the city by walking around and talking about anything at all with anyone I chanced to meet. Instead, I was rushed from teas to dinners, bored to extinction by the parochialism of the talk. But at least I was taken to a session of the Senate so I could observe American democracy in action, and I even succeeded in snatching some unforgettable moments in the wonderful Library of Congress. But I had had enough and cut my stay short, happy to get back to the anonymous crowds of New York and to my husband, who was waiting for me there.

Back in France, I realized that from then on I would have to live with electrical coring: Day and night, its lines and curves were stamped on my existence. Henri, jubilantly optimistic, wanted apparatus and engineers. He went so far as to think that petroleum people were holding their breath, waiting to see what he could do next. Young, his life before him, he exulted at the prospect.

8

A Contract with the Soviet Union

IT was 1928. Vahe Melikian, a young Russian graduate of the Ecole des Mines in Paris, introduced himself at the Rue Saint-Dominique. He was interested in geophysics.

A native of the Caucasus, he had fine almond eyes, olive skin, and silky black hair. He was of a pensive, rather withdrawn temperament; he seemed to brood over his mysteries. But, like it or not, he could not make a secret of his intelligence: It sparkled.

A year passed, during which Melikian studied methods of electrical prospecting, won the confidence of those over him, and adopted the ways of the company. "Don't you see, Monsieur Conrad . . .?" Conrad did more than see: He felt a growing friendship. Marcel, maintaining his reserve, did not admit to himself that he too was feeling the ascendancy of the young Russian. Then, without saying a word, Melikian brought a solution — unexpected, to say the least — to the financial difficulties and uncertain future of the business in the United States. Had he this solution in mind before he came to work? If so, he had not let anyone know.

One day, in March 1929, he presented himself at the office of Conrad and Marcel, accompanied by an official. When the door was closed the stranger revealed his identity as Professor Golubiatnikov, representing Soviet geology. He had come to propose to Messieurs Schlumberger a contract setting up a close working relationship in return for regular payments from the U.S.S.R. Melikian, the young Russian engineer, would be part of the contract: Back in the Soviet Union he would serve as liaison between the Soviet oil trust and The Société de Prospection Electrique. Was this proposal agreeable to Melikian? His expressionless face gave no hint of his thoughts. As for Messieurs Schlumberger, they were to commit themselves to keeping secret the results of their work, as well as other information they might have occasion to pick up in the U.S.S.R. They were also asked to stay clear of all financial, banking, and political groups, not to intervene in internal Russian matters, and to exercise only their functions as experts. They were to publish nothing without the consent of the Soviet trust, and any new discovery was to be made known to Moscow at once.

These conditions might be dangerous and had to be thought about. Might not the results obtained in the U.S.S.R. filter across the frontier and be seized upon by *Pros*'s competitors? Would the Russian engineers observe professional secrecy? Since the Revolution, contacts with Soviet Russia had been few and far between. Would closer collaboration be regarded as an open approval of the new regime? Would it not compromise possible future dealings with the U.S.? Finally, might not the good name of the family be jeopardized? On the other hand, by refusing the contract, a really unique opportunity might be missed. There was more than money involved: For the first time their research and prospecting methods had met with confidence and acceptance without the cautionary proviso of a successful tryout on the ground. These people had enough imagination to realize that the proof of a work was the work itself. They did not say, "Show us first and then we'll think about it"; they said, "Come along, we're putting our money on you."

Conrad paced the floor, his thumbs hooked in his suspenders, his head bent forward. He saw a new world opening before him, a world in which comradeship and confidence would regulate the relations of each for the good of all. Marcel, seated, smoking, listened to his brother and read the contract over and over again: This was a serious moment, not the time for flights of idealism. What unknowns were concealed in this commitment? Would not partnership with a State agency hobble his freedom and mortgage his future and that of the engineers he would take along with him? Would these engineers understand that a single imprudent word or unsuitable attitude could have unforeseeable consequences? Conrad foresaw hardworking teams of young men united by a shared enthusiasm; he liked to think of this bond. Marcel, skeptical, signed.

The first five engineers chosen for the U.S.S.R. boarded the ship *Le Rhin* at Marseilles in August 1929. A violent mistral* was blowing; the ship tried three times to get out of the port. "Come on, Skipper, where's your nerve?" shouted the pilot, clinging to the wheel.

*A very strong, cold, dry wind in Provence (southern France).

It took a long time for the story of the voyage to get back to Paris. A wild sea laid hold of them as soon as they were out of sheltered water. The captain of the freighter had the heart of a pirate, and he had a pickup crew. The ship's cook, formerly a bird catcher, seasoned his dishes with canary seed; the doctor, a toothless opium addict, had neither tools nor title; a seaman had his thumb crushed and the "doctor" cut away the flesh with nail scissors. After several days of stormy weather the ship tied up at Mudanya to unload pipe, which men carried off on their backs. It was a long process and the travelers, happy to forget the freighter, had time to make a leisurely tour of the city of Bursa and its green mosque. At Odessa, a guard came aboard and forbade them to set foot on land. The weather was hot, and the passengers wanted to get ashore to the beaches; but they were told that they would need a special permit. After two days of parleying they had their permit, but the weather had changed and bathing no longer looked attractive.

Three weeks on board finally brought them to Batum. The representative of the Compagnie Paquet slipped aboard like a hunted animal and never stopped saying, "Hide me, I want to get out of U.S.S.R." The captain addressed a stream of predictions to the engineers as they prepared to go ashore; he thought they were crazy to go to work in such a country.

The train rolled along slowly from Batum to Tiflis, and the beauty of the landscape took their minds off their uneasiness. At Tiflis, Melikian, whom they had known in Paris, met them on the station platform. He seemed changed—distant, more withdrawn, keeping his voice low. There they were: five young men eager to understand, yet not daring to give vent to their curiosity. The taciturn reserve of their former comrade, now their chief, silenced them. He was a shadow of the man they had known, yet he seemed free in his movements. What had happened to him in the past weeks? To their mute questions his impassive face was response enough: "Don't ask me, learn for yourselves." Surprised and disquieted, each man kept his peace.

The trip across the Caucasus on a road full of potholes brought their travels to an end. Were they—like the knights bound homeward from Jerusalem—about to be drawn in by Queen Tamara, whose castle, built like a guard post over the road to Vladikavkaz, dominated the wild gorges of the Terek River? Tamara, a hospitable queen, offered shelter to the travelers who passed below her ramparts. She also made lovers for a night of men who attracted her, and her passion would not allow her lovers ever to belong to anyone else; so, the morning after, she had them thrown from the top of the tower. However, nothing was left now of the queen's castle but some ruins and the horror awakened by a poem of Lermontov.

Grozny, a Chechen city, was an agricultural center where several hardly passable roads (always either deep in mud or deep in dust) converged. It was also "oil country." The kulaks, who had been evicted from their fields and houses, lived in the earth in dugouts; only a wisp of smoke rising here and there indicated a kind of larval life beneath. The young engineers — the men of the day, the petroleum aristocrats — were housed in a cement building that looked like a barracks. This was headquarters for the trust's administration. Meals were taken in a room called the "club"; the Russians and the French were separated. Since "free" food was not to be found and the local fare was indigestible for foreigners, a special diet was provided for them. If the Russians resented this, they did not show it, and as for the French . . . well, everybody knew: good input, good output . . .

On the street or in the field, they were the focus of a discreet but lively curiosity. The young Soviet engineers were enthusiastic and eager to learn. It was not easy for their French colleagues to prospect and to teach all at once: Language constituted an almost impossible barrier and they had only one interpreter for the six crews. So they quickly set themselves to learn Russian.

Each man had his own system. Lannuzel stuck to his grammar book; he had it with him all the day long, and at night he made it his bedside book. Bordat lost no time finding a Russian woman: "The method has more life," he said. Poirault, the group's intellectual, tried exchanging ideas; but, with his limited vocabulary, he ran no risk of being taken for an expert in Marxist philosophy. Sauvage pointed his finger at things, asked what they were called, and repeated the words like a child until he knew them by heart. Being a nature lover, he was the first to be able to name the various trees and plants. Jost, who by temperament was inclined to adopt the Bordat system, intended to look before he leaped and to follow the results achieved by the others. He also relied on authority: "Monsieur Conrad will certainly advise us concerning that method," he said, with a tiny, expectant smile.

Everyone was awaiting Conrad's arrival. The heads of the trust were uneasy about the "Professor," the savant who would be able to solve all the problems. Well, he was on his way, he was coming — he, the man they talked about as if he were equipped with a magic wand.

Conrad, impatient to discover the "socialist" world as he imagined it, landed in Moscow in November 1931. His young Russian friend welcomed him with a rather stiff deference that was not at all usual in their relations. But how could one refrain from asking questions, from interrogating the faces of the people who looked at you as if veiled by an omnipresent *nyet?* Outside the station, a comfortable American car was waiting for them. Melikian, saying nothing, took his seat beside Conrad. The car plunged through the crowd; they rode through Moscow at high speed without exchanging a word. In a mansion at 12 Leninskaya Street a very large, richly decorated room had been made ready for the "Professor." The Professor was not pleased by this spurious ostentation. He had come to Russia to see what a people's revolution had accomplished, not to be treated like a prince out of the past.

After several days of receptions and conferences Conrad still felt a certain uneasiness. Was he going to be disappointed? Why be surprised that an inventor, even a modest one, received support and encouragement here? While his work belonged to everybody, at least he received credit for it in his lifetime, not posthumously, as was too often the case elsewhere. Every creative effort was in itself a small revolution: What wonder, then, that the country of the Revolution recognized its own?

But he was not fooling himself, and he smiled a little at the turn his thoughts were taking. He was a solitary worker whose family were the only ones who approved what he did; the honors being paid him in the U.S.S.R. were beginning to go to his head. Conrad's first contacts had indeed been most cordial. He was impressed by the high degree of interest he encountered and by the enthusiasm that scientific and technical progress stimulated. One did not need to be a scholar in order to see that this people had the dynamism and ardor of youth.

The "Professor" left Moscow in a comfortable sleeping car draped in red velvet; his young friend, who served him as interpreter and as a screen between reality and myth, accompanied him. But, although Conrad was not necessarily hostile to myths, he was quite able to see things as they were. Maybe bureaucrats, technocrats, and the military did not constitute a propertied class in the traditional sense of the word, but they certainly belonged to a privileged caste. It was clear to him that the level of life of this new elite far surpassed that of the masses. Proof enough was this luxurious sleeper, which was coupled to the end of a train made up of cars with wooden benches and crowded with shabbily dressed travelers. Melikian, sitting beside him, did his best to turn Conrad's thoughts away from political and social problems. What seemed to interest the Russian was the work of Gide and its influence on French youth: *L'Immoraliste, les Nourritures Terrestres.* . . . Ah, he seemed at ease in handling ideas foreign to his own country.

The Russian plain stretched endlessly. Conrad looked out over the monotonous landscape, where groups of buildings rose here and there — the kolkhozes. He could see peasants working under the watchful eyes of guards with bayonets fixed on their rifles. Prolonged stops in the stations — often due to mechanical failures — gave him an opportunity to stretch his legs. Yet Melikian never let Conrad get out of his sight; these halts, which were to Conrad a source of surprise, made the Russian uneasy. The stations teemed with people waiting for the train. Sitting or lying on the bare ground, almost stacked on top of each other, they gave off a strange odor of old leather. In the dining cars, unnameable leavings on dubious tablecloths awaited the courageous client. After three days and nights of asking his companion questions, sipping weak tea, looking at the plains, and strolling in stations jammed with waiting crowds, Conrad was glad to reach his destination.

Grozny is situated at the foot of the Caucasus range, in a vast plain where the vegetation withers under the summer sun. The arrival of the "Professor" was a major event in this dusty town. A reporter, anxious to have his story carried in *The Grozny Worker*, was at the station to interview him. What was the Professor's opinion on the subject of Soviet petroleum engineers? The Professor's opinion was that young Soviet engineers were in too much of a hurry; they wanted quick results when what was needed was patience and tenacity. Was it a bad thing to want quick, immediate results? the reporter asked worriedly. Electrical coring was in its infancy, the professor replied; beyond its immediate application in the field and in the boreholes, there were numerous scientific studies to be made in the laboratory. But, the journalist went on, now that the Professor was there in person, Soviet petroleum would.... The "Professor" smiled understandingly: "The program is an important one, it requires a huge technical effort which calls for wholehearted collaboration from your geological service concerned with drilling."

The house occupied by the administration of the trust was ugly inside and out, but it offered comfort and pleasant hospitality. In the lodgings reserved for them, the French engineers prepared a feast to welcome Monsieur Conrad. They all surrounded him at this first meal; one would have thought it was a happy family getting together. Conrad encouraged them to adapt to the habits of the region and to learn Russian; he questioned them about personal problems and about the technical difficulties that each of them was meeting in his work.

"Let's see, Lannuzel," he said, "you're just back from the steppe. Tell us about your life there." Lannuzel did not wait to be asked a second time: His surface crew went around all day long in the heat, made a lot of measurements, and ate a lot of dust (in fact, to hear him, dust was their principal food). Four weeks of this nomadic life in tents had transformed their return to Grozny into a descent into a veritable dreamworld.

Sauvage, for his part, had a long story to tell. Shortly after his arrival in Grozny, one of the directors of the trust had offered to show him around the area where he would be working; they were to meet at his office the next morning. Sauvage was waiting for him at the appointed time when an employee came to tell him that the director had gone off on vacation. Sauvage did not believe this and pressed the issue, whereupon the other man, visibly upset, made himself scarce. It came out later that the unfortunate director had been arrested the night before. His successor was a man of decision. "I count on you to increase production; I've got to have oil," he announced right off. His voice had the energy of despair.

For weeks Sauvage tried and tried again, but in vain: The fact that the boreholes were deviated prevented the coring sonde from reaching the desired depth. The director's life hung at the end of the sonde, the future of the *piatiletka* (Five-Year Plan) depended on it; but the sonde stuck halfway down. Sauvage dropped by the director's office again, with its forest of black numbers covering the whitewashed walls and the production curves conforming to the Five-Year Plan. . . . Sad to say, neither figures nor curves could help a bit—the sonde still would not go down. Sauvage, exhausted, came back to Grozny. The man with the voice of decision was accused of sabotage and disappeared. His replacement was a fatalistic sort. "You see that armchair?" he said. "It doesn't bring good luck." Sitting on an ordinary straight chair, he stared at the empty armchair that had swallowed his predecessors, thinking: "Petroleum? We don't believe in it; it's one of those things people talk about, an abstraction they die for, but nobody has ever seen it."

Sauvage listened, astonished at such frankness. A Soviet citizen never confided in a foreigner—no one dared, so closely was the solitude of each one watched over by the solitude of all. Was this man trying to impress him by confidences of this kind, as if heart could make up for deficiencies in technology? Pensive, Sauvage returned to the work site. Would this heretic who did not believe in the reality of petroleum—except that it caused ill fortune—have any better luck than his colleagues?

After several days of effort, the sonde began to descend. The apparatus recorded measurements; then, sure enough, at a certain depth. . . .

Sauvage could not believe his eyes; he tried the next borehole, spent the night on the job, and obtained very promising recordings. Impatient, he hurried into Grozny and burst into the director's office. The diagram was snatched out of his hands; each person wanted to interpret it then and there. There it was, in black and white: a deposit of oil, maybe a new pay load. That day the trust people could relax.

Bordat had only a short adventure to relate. He had been going over a field of corn, and dragging the cable had not been doing the crops any good. The Chechen, an agrarian people with no love for the Soviets, were disposed to massacre anyone imprudent enough to venture into their fields. One of them started after Bordat, who pictured himself slain and abandoned among the cornstalks. The proximity of the truck saved his life; the incident left him with a curious, poetic memory. The Chechen are always ready to kill each other in obedience to the law of retaliation: Death calls for death, vengeance demands vengeance, and this goes on forever. Lest the law be forgotten, a tall pole is planted over the grave of a person who has been killed. Once the deceased has been avenged, the pole is simply moved to the new grave.

Conrad listened. He watched these young men who were so enthusiastic about the work that had launched them into a life with an uncertain future. Tomorrow they would have lived that life; the day after tomorrow they would be old. Would they bear him ill will for having uprooted them and made them vagabonds at the four corners of the earth? For the first time, he caught a glimpse of the possible consequences of what he was doing—fathers and sons separated, families transplanted, perhaps his own as well. . . . He had never imagined that his modest laboratory and his solitary research would one day grow into something of which he would no longer be master.

Conrad Schlumberger in the basement of the Ecole des Mines (1912).

The first truck with its equipment, in 1912, in France.

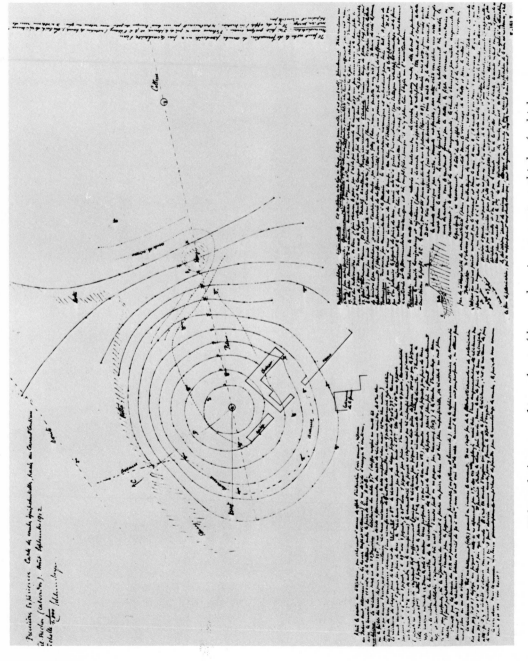

"I have before me a large sheet of heavy drawing paper, two-thirds of which are covered by roughly circular lines . . ." —Conrad Schlumberger.

A wooden tower and its pipes in 1925. These towers were seen in oil fields all over the world, from Romania to Péchelbronn and America. In the Los Angeles area they stood side by side like trees in a dense forest.

Conrad Schlumberger taking a measurement in the field, Normandy, 1912.

The famous "black box" used for surface measurements, not for electrical logging (photo by Michaelides).

Marcel Schlumberger, on the right, in the laboratory at Vitré, Brittany, about 1928.

A typical scene in Louisiana, 1929, showing Schlumberger truck with winch and tool chest. Engineers and assistants often spent hours clearing the mud off the truck before starting to take electrical measurements.

A parasol to protect men and apparatus in the California desert, 1932.

Jean Mathieri and his truck, in the mud of Texas, 1932.

Henri Doll taking measurements in the desert around Bakersfield, California, 1928.

In 1932, California: Legrand and Deschâtre taking measurements in a borehole with Schlumberger equipment.

*"We need oil or we lose
our jobs," said the Russian
oil men in 1930. Electrical
measurements meant life
or death then!*

"Living and working conditions were primitive.... The roads were impassable; to go from one place to another with the equipment, an oxcart was often the only vehicle that could get through the winter mud." This photograph, taken in Romania in 1932, shows a team of horses pulling a truck out of the mud.

These barges are ready to be moved to the site of a swamp in Louisiana where an engineer will take electrical measurements, circa 1953.

Jean Schlumberger and the author at Braffy in Normandy.

9

Eastward Ho!

CONRAD owed this Russian venture (which might be the beginning of a worldwide enterprise) to Vahe Melikian, the young companion whose presence had become a necessity. When he did not find the Russian close by his side, the absence surprised him. The friend one questions, the friend whose silences are never refusals. . . . Conrad knew that Melikian was different from the other Russians—different, for instance, from this Glutchko, the head geologist, with an oversized head perched on a puny body, a face with prominent cheekbones, and restless eyes in search of protection.

Dressed in a tight jacket that buttoned up to the chin, Glutchko paraded—in well-polished shoes, as became a top man—up and down a large, sparsely furnished room. He waited until Conrad was seated, befitting a person of importance. He unlocked a drawer, took out his supply of carefully wrapped sugar, chose a very small piece, placed it on his tongue, and sipped some straw-colored tea from the saucer. Having carried out this rite, he felt at ease.

"The diagram shows oil, doesn't it, Professor?"

Conrad looked at the long strip of paper; zigzagging lines traced the structures in the subsurface. "It's difficult to be sure," he answered. "I would prefer to run the trials again."

But to Glutchko, the strips and their diagrams were like the perforated rolls of a player piano; they gave off chords and harmonies. *Let the Professor quibble all he likes*, he thought, *we've found a productive stratum: We'll stay on as director for many fat years to come.* . . .

Melikian's friendly, silent figure appeared in the long corridor. "Well, well! I didn't know you were there," said Conrad.

"Monsieur Conrad, Professor Grigoriev is waiting for you."

"Professor Grigoriev?"

"Yes, a former Orthodox priest. The events of October changed his course."

Grigoriev was a tall man with an imposing air, very much of the *ancien régime*. He could not hide it — his fine face, his neatly trimmed goatee, his military haircut gave him away. His French was as stilted as his person; his florid form of address smacked of the Academy. "Monsieur le professeur, allow me to tell you that your discoveries enrich science in general and geophysics in particular, while your presence in the Union of Soviet Socialist Republics honors . . ."

As for honors, Conrad wanted no more of them; there were far too many already. He handed Grigoriev a diagram: "I would like to know whether in your opinion this curve . . ."

Grigoriev had no opinion — or if he had, he did not give it (or else he waited to give it only after he had read it in *Pravda*).

A young Russian engineer entered the room, waving the last diagram of the day like a flag: "Congratulations! You are the revolutionary of geophysics!" (Conrad bent his head; he no longer knew exactly *what* he was.) "Here is an abandoned field which, thanks to you, will produce. Professor, I am betting my future and my life on your methods!"

The "Professor" shook his head. "We shall do the best we can, but I would rather have you not risk your life, nor even your future."

The other man turned a deaf ear to this. The diagrams, electrical coring, and the new methods would make him the big man of Soviet petroleum. Better to go prospecting in the forests of Siberia than to gather moss in a subordinate post. . . . Conrad cast an alarmed eye upon the young enthusiast, tried to calm his zeal, then finally gave up and let him go on as before.

The news of the success at Grozny had spread to Baku, and the town prepared its reception of the Professor. The day of his arrival was a gala occasion — crowds, bouquets, red carpet and all. Zametov, the geologist, his beard waving in the wind, bowed ceremoniously. Conrad, dumbfounded by all this show, sought some relief for his timidity in the young, candid faces, often of simple workmen, around him. While the compliments, expressed by smiles and hand-waving, went on and on, he met the green-eyed gaze of his silent companion . . .

In Baku in the distant past, there had been a small boy in short pants, white stockings, and black buttoned shoes. A high, starched collar framed his melancholy face. Seated on a hassock covered in pink satin, he looked out the window at the clouds of sand falling on the passersby. The glass of tea and the pastries on a silver platter went unnoticed. Was Melikian thinking about this other self of his? Did he feel a part of this somewhat ridiculous pomp and ceremony, or was he dreaming of images of the past? . . . "Not at all, Monsieur Conrad; for me the past no longer exists."

The noise around them notwithstanding, a deep silence enveloped them.

To honor the Professor, a banquet was given the evening of his arrival. The area bigwigs and first-line workers had been invited. Pink and green garlands decorated the hall, reminding Conrad of Prize Day at the little elementary school in his Alsatian village. Although he had not been asked to make a formal speech (required by Western custom), all those present were eager to hear from him; so he thought it well to say a few words, Melikian translating as he spoke. As he saw things, Conrad said, what struck him as a foreigner most forcibly was the effort directed toward increasing the country's economic potential. The petroleum industry's program, one of the most ambitious anywhere, was on a scale with the immense riches — as yet scarcely explored — of the Soviet subsurface. The oil trusts had understood the importance of modern methods in geophysics. He was happy to observe that teams of Soviet geophysicists were thinking of devoting themselves to lines of work that he himself had staked out; and since he had just referred to his own research, he added that electrical prospecting of different terrains required a long and arduous apprenticeship. This was because between the reading of a diagram and the actual finding and exploitation of a bed of petroleum there was all the difference that separates ideas from practice. Conrad closed by adding (a bit pointedly) that if perhaps he was a "learned professor," he was nonetheless what was called a capitalist, and one grateful indeed to the Soviet Republic for the confidence it had been kind enough to extend to him. Whereupon, there was prolonged applause, and the vodka flowed in torrents.

From the five engineers originally sent to the U.S.S.R., the number had risen to fourteen. One of them was comfortably settled in Moscow, another was condemned to the icy cold of Lake Baikal, still others ranged over the grassy steppe. One almost drowned in the Caspian Sea, while another, for lack of anything better, fed himself exclusively on caviar. In the deserts of Kazakhstan, they lodged in round huts roofed with thick felt to keep out the sandstorms. In summer, in order to ward off the clouds of mosquitoes whose buzzing made it hard to hear one another, they burned camel dung; in winter, in order to ward off the cold, they still burned camel dung. But what most sorely vexed one of them, whose work brought him into such solitude, was that right there, under his nose, 200 miles from the nearest settlement, he had been robbed of his business suit and his underwear. In a letter that took ten weeks to reach its destination, he besought the Paris office to act as intermediary between him and his tailor; and—since even at the farthest reaches of the desert one was not sure of keeping one's socks—he wanted to know whether his insurance covered the risks created by wild beasts, the elements, hunger, thirst, frostbite, and a number of other calamities. In the answer he received—it was worthy of his questionnaire—he was informed that Paris offered him 2,000 francs as indemnity for the loss of his wardrobe, that he was insured to the amount of 200,000 francs in case of death or total disability, and that this insurance covered all accidents including rabies and black lung; the only exceptions were death due to dirigibles and other airships or due to engaging in violent sports.

The truth is that in these countries death was looked upon differently than it was on the banks of the Seine. One of our engineers, stranded among the Cherkess, had seen old men drag themselves to the burying ground, lie down in their graves, and wait for death. To die alone and with dignity was the supreme act that crowned existence. In the same village, a Circassian woman with long blond braids, encased in an iron corset, waited for her fiancé to set her free publicly, in a kind of ritual abduction, on the day of the wedding. Indifferent to the giant derricks that were pumping the future out of a geologic past, immemorial customs regulated the march of time.

O unknown friends, I imagine your struggles, your hopes and disappointments. How often I dreamed of coming to visit you! . . . Then, one day, in January 1932, my husband decided to go; I prepared to accompany him. To his baggage—apparatus, cables, and quilted clothing—I added my own: a notebook with a glazed black, supple cover in which to record my impressions. To go to the U.S.S.R. was like tasting forbidden fruit. As I waited for the hour of departure I was curious, but without the slightest preconceived idea. Only my notebook, whose glossy cover I caressed to prepare it for my confidences, knew my joy.

Traveling across Europe is a long and tedious experience, even with a notebook at the ready; but to make up for that, what style, what luxury in Moscow! Receptions, evenings at the Bolshoi, walks along the banks of the Moskva and around the outskirts of the Kremlin, kasha and zakouski, vodka and more vodka. There would never be enough pages in my notebook to hold such a succession of marvels!

I had filled an uncalled-for number of leaves in my desire to describe a truly extraordinary operetta: At the gates of a paradise obviously conforming to Soviet dogma, we saw St. Peter refuse entrance to a potbellied priest but welcoming a scarlet woman to the sound of trumpets. I left the play all bewildered, wondering if I was the only one who saw the comic side of the situation: If a citizeness as unproductive as a prostitute merited reward — and certainly a reward was intended, since the priest had no right to it — then why deprive oneself of everything in the name of "socialist" productivity? Yes, I may have been alone in perceiving this comic aspect — I, whose deluxe boots dug into the soft snow and who slept in a very large bed inhabited at one time, perhaps, by the shade of Rasputin, in the Yusupov palace. *Tomorrow*, I thought, *just to set matters straight, I shall go and look at Lenin in his mausoleum in Red Square: I hear that his embalmed body has kept its natural smile.*

The journey from Moscow to Grozny seemed endless. We had not thought to provide ourselves with food and drink. My travel notebook shows this: When one's stomach is in one's boots, one is not much given to confidences. But oh, the pages filled with Grozny! The market, so Oriental in appearance, offered shriveled apples, rusty nails, used buttons, bits of string in small piles on the ground. A flea market in the fullest sense of the word, where nothing was left but the fleas and the handsome Caucasians, strong and disturbing with their scimitars at their side. At night, their shadows marched across the whitewashed walls of my bedroom and kept me awake. I wrote this in my notebook, although I did not dare talk about it.

At Baku, clouds of sand stung the pages of my notebook. Instead of filling it with random impressions, was it not time to choose, to take sides? The Russia of the Soviets, a new world, a world apart, demanded adherence or rejection. The undecided were not welcome, nor were skeptics, nor women of my kind, who asked more questions than was allowable. But how could one not question the girl who, with hatred in her eyes, snatched a bouquet of narcissus out of my hands? Or the former diplomat who was supposed to be giving me Russian lessons but talked only about mass deportations? Or the schoolteacher who was installed in one corner of a bedroom, rolling out his mattress on the floor in the evening, while a couple did likewise in the opposite corner? "What's one lifetime?" he replied to my questions. "There will be other people, and they will enjoy social progress."

Yes, what is one man's lifetime? In the oil fields I saw a workman give up his own meal — a meager roll of rice and meat — and offer it to the French engineer. As a sign of what? Of gratitude, of thanks for "social progress" in which a human life is never thought of as anything but the future of someone else, the future of these others who "will be"?

In the big, bright pink building, engineers and workmen mingled in a bluff camaraderie that I found appealing. I went from one to another, naive enough to put my trust in questions. "Comrades," I would say, "What is the life of a man, *pazhalista*" [if you please]?

Doubts, uneasiness of mind . . . these were "clandestine articles" prohibited within the U.S.S.R., shameful things not to be displayed in public—above all, not to be exported. Sad to say, my precious notebook, faithful witness of my emotions, was confiscated at the border by a customs officer full of "socialist" zeal. I missed those notes terribly; I had counted on rereading them in the hope of finding some light on the eight weeks I had spent in the U.S.S.R. It took me a long time to see how ironic was the confiscation of my notebook: What else was it but the beginning of an answer to my questions?

Back in France, I had the feeling that I was lost in a labyrinth. Every turn beckoned me on, then immediately put me off. It was as if I were learning my sense of direction all over again, just beginning to rediscover my freedom of movement.

As they did in the Soviet Union, so at the Paris laboratories young Frenchmen were initiating their Russian comrades into the secrets of electrical prospecting. Seven years of continuous exchange were drawing to a close. The days were gone when the "Professor," welcomed to the Academy of Science in Leningrad, was borne aloft in triumph. At the Rue Saint-Dominique, Conrad was thinking aloud in the presence of his Russian counterpart, who said not a word. What did his silence conceal? Did an ex-driller, now promoted to the rank of top man, know more than Conrad about what was coming?

"Seven years," said Conrad. "Seven years of good work, during which the Soviet engineers worked hand in hand with our men. The French learned to speak Russian, they put forth their best efforts to give satisfaction to the oil trusts . . ."

A smile that said nothing curved the lips of the man who was listening. Dakhnov, a sentimental, passionate man who was doing a period of training, had also joined them.

"In 1932," Conrad continued his monologue, "a second contract extended our collaboration; then came a third in 1934. Last September, we were talking about a fourth contract dealing with the seismic method. During these several years, we have prospected together in Central Asia, on the shores of Lake Baikal, in the Sakhalin region, in

Kamchatka. Now, all of a sudden, communications with Moscow seem to be blocked."

The blond, rosy, younger Russian started to protest: Hey, wait a minute, how about him, what about his training in Paris if ever. . . . His compatriot's frozen smile cut him off—Conrad, too, for that matter. He blamed himself for having spoken openly to this bureaucrat, smug as he was in his commanding wordlessness.

A few days later, Mikhail Opochinin, a former White Russian whose knowledge of French had earned him a subordinate job at *Pros,* laid his weekly summary of *Izvestia* and *Pravda* on Conrad's desk. "It's the *chistka,*" he announced triumphantly. "Purges and trials right down the line."

My father's incredulous look spurred him on. "The ones who go to trial are lucky: They get their names in the papers. The rest will disappear without a trace. You can see for yourself, Monsieur Conrad. Sabotage, sabotage: It's the only word the commissars know . . ."

Conrad read and reread the typed pages. How was one to understand exactly what was going on? What did this hermetic jargon mean? What was the key to it? . . . Or else, having staked so much on Soviet youth, on their efforts and their enthusiasm, was he not deliberately refusing to read between the lines? In France, the Communists were playing the "hands-of-friendship" game and in Russia . . . "Leave them alone," people around him said. "Let them tear each other to pieces. There's nothing you can do about it."

Nothing? Not even for those who, side by side with the French crews, were probing the Russian steppe for the greater good of Soviet petroleum? After all, Conrad had some weight in Moscow; hadn't he just subscribed 40,000 rubles for the second Five-Year-Plan—22,000 from the Société de Prospection Electrique, 18,000 in his own name? Let things alone, put up with his own pain? He couldn't. In mid-March, he decided to go back to Russia.

Vahe Melikian, emaciated, a shadow of his old self, was there in the half-light of the station to meet Conrad as he got off the train. He was not talkative by nature; now his one-word replies echoed each other.

"Zametov?" Conrad asked.—"Arrested."—"Glutchko?"—"Arrested."—"Grigoriev?"—"Arrested."

That night Conrad got no sleep. These men, yesterday at the head of the oil trusts and today vanished—what crimes did they have to answer for? And he, the *Frantsuski Professor,* what in the world could he hope to accomplish? To fight? But against whom? Against what?

What could he do in the face of an ideology he had thought he understood but understood no longer? Did he think he might succeed in saving the life of one man, if only one? Wasn't it beyond him? Then why had he come? Was it, perchance, in spite of himself, to negotiate a new contract?

The following day, accompanied by Melikian, Conrad walked across Novoguine Square, which was dominated by the building where the Institute for Geological Petroleum Research had its offices. The low houses around the square were in bad repair. Streetcars went rocking and clanging in all directions; dark bunches of passengers hung on for dear life, a red scarf adding a brave note here and there. Pedestrian traffic was heavy.

"Your newspapers," said Conrad, as if taking up an interrupted conversation, "are talking about a turning point in history, about socialism fulfilled at last. When I look around me, I have a hard time believing it."

Melikian's only reply was, "Here we are. You will be received by Golubiatnikov. I doubt that you will recognize him." And, as they mounted the stairs, "Don't be surprised: He has aged twenty years."

Golubiatnikov was waiting for them on the geophysics floor. Nothing was left of his self-assurance or his jovial voice. He might have been seeing the foreign "Professor," whose "revolutionary methods" he had extolled in the past, for the first time. A void surrounded him; his armchair floated in the void, and if he still functioned at all, if he still even wiped his glasses, it was as if by mere habit.

Melikian translated. Conrad had the impression that Golubiatnikov either no longer knew what was in his files or was pretending not to know. He evaded, he minced words; he seemed on the point of asking why the Professor had come to Moscow.

Other meetings followed—with him and with his peers—men highly placed in Soviet geophysics, now reduced to the role of puppets. To express an opinion or to make a decision was no longer within their province. A detail settled or a point agreed upon in the morning was questioned again in the evening. The same words changed meaning from one sentence to the next.

Several weeks of this game (or of this systematized duplicity, Conrad was not sure which) left him exhausted. Finally, a kind of agreement was reached, by the terms of which the French would function thereafter in the Soviet Union only as technical advisers, principally for local manufacture of Schlumberger equipment. At least, Conrad thought, he had obtained some assurance that relations would not be definitely broken off and that Melikian would continue in his post as liaison agent between Paris and the Soviet oil trusts. Alas! Melikian was not to escape the Stalinist reign of terror, and Conrad, who thought he had not entirely failed in his project, took away nothing but disappointment and impaired health.

His departure from Moscow touched Conrad deeply. He left from the same station where some weeks before he had arrived feeling some confidence. The man who had welcomed him stood on the platform. Did they sense that they would not see each other again? But the train began to move, and the last look they exchanged put all the world's past between them. The present? They rejected it. The future? They no longer believed in it.

At Leningrad, Conrad boarded a ship for Sweden. "I will tell you all about these painful weeks," he wrote to me.

Too late. In Stockholm, in the office of a business lawyer, he collapsed.

I saw him in a charming home for old ladies, to which he had been carried. Eyes closed, unable to speak, he no longer could express himself except with his hands. "Don't worry, I'll hang on," his heart murmured. But his breathing gradually failed, and so his soul left us. My soul fell silent; for a long time it was as though asleep.

This was in May 1936. The weather was mild for the season. The train that carried Conrad back to France traversed great birch forests. On that final journey, he carried away a handful of Swedish earth; I left the best of myself behind.

In Normandy, in the shade of a little church, he was laid to rest. The son had rejoined the father.

> *Le sapin sera planche*
> *Et planche de mon lit*
> *Quand je serai ta branche*
> *O forêt de la nuit.**

In Paris, at the Rue Saint-Dominique, life began again. In thought and in word, everyone behaved as if Conrad were about to appear on the threshold of his office. Marcel made him part of his work: "My brother thought. . . . My brother wanted to know . . ." Among the staff,

---

*The sapling will be plank
 And plank of my bed
 When I shall be thy branch
 O forest of the night.

from the stenographer to engineer, the habit of invoking his presence continued for a long time: "Monsieur Conrad told us. . . ." It was the war and the German Occupation that made Conrad's name slip into the shadows.

Between Moscow and Paris, what was left of his last efforts? Promises and more promises, all wrapped in subterfuge — a contract in its agony. By a miracle, Vahe Melikian managed to be lent to *Pros* for six months . . . six months spent counting the days. During the week preceding his departure, we no longer dared look at him. Melikian was married to a Frenchwoman and was the father of an eight-year-old son; he could have stayed, he should have stayed. But he was a man of his word, and left France at the appointed hour. His farewell was that of a condemned man. From Moscow, he wrote: "The only way to maintain relations is to accept silence." In the U.S.S.R., from the banks of the Bug to the shores of Sakhalin, terror reigned. February 1938 was the last time we heard from our friend.

In 1971, thirty-five years after my father's death, on the occasion of a petroleum convention at the Gubkin Institute in Moscow, Charrin and Jost* were surprised to see, hung between the portraits of two Soviet geophysicists, the portrait of Conrad Schlumberger.

By way of a footnote to history, I may add that at its beginning in the U.S.S.R., electrical prospecting was frankly and simply called *adine* (one) Schlumberger.

*Two French engineers.

10

Development and Decline in the U.S.S.R.

PROS was among the first French business firms to establish technical and commercial relations with the Soviets. These relations extended over seven consecutive years, dating from the first contact on July 4, 1929, to the summer of 1936 (when the infamous Moscow trials began), and were generally cordial. It can be said that *Pros* had, in a way, granted favored-nation status to the Soviet oil trusts. Whereas the *Pros* rule was to offer services performed from beginning to end by French personnel and with French equipment, the agreements with the Soviet Union included not only petroleum prospecting but also the manufacture in the U.S.S.R. of various apparatus and the training of Soviet engineers both there and in France. This modification of *Pros's* regular method — the tendency at *Pros* being to guard industrial secrets at all costs — involved, as I have noted, a risky revision. To familiarize the Russians with its methods, inventions, know-how, and techniques — wasn't this like tempting the devil? Of course, the contracts contained protective clauses, but if the occasion arose, what could a small private enterprise do when up against a state monopoly?

But would *Pros* have run the same risks in a less precarious economic situation? I am not sure. For instance, in the United States, *Pros* took no such chances, although Schlumberger engineers were working there until late in 1930, when the last of them were withdrawn.

However, that may be, fears and hesitations turned out to be without much foundation. A few articles appeared in the Soviet press, accompanied by the usual trumpetings about the "effectiveness of Bolshevik work." *L'Humanité*, the French Communist daily, mentioned the young French engineers who had gone to the U.S.S.R. to work for "the buildup of Socialism." A book written by the same Dakhnof who had trained at *Pros* in 1934 was brought out in Moscow by the Nefteizdat publishing house; it was summarized in *Geophysical Abstracts* (August 1935) as "a detailed description of the Schlumberger theory, with considerable practical material on electrical prospecting." But on the whole, the Russians' discretion was satisfactory. I remember that every *Pros* engineer committed himself in writing:

> " . . . not to make use of knowledge or impressions which he may gather during his stay in the U.S.S.R. in the course of his work, in order later to publish books, articles or studies describing life in the U.S.S.R. or analyzing the social and political organization of the Union of Soviet Socialist Republics.

The Soviets had bought silence along with methods of electrical prospecting.

Nothing — or next to nothing — leaked out about the real working conditions in a country that flattered itself on having abolished capitalist exploitation. There were no immigrant workers in the Soviet Union who were free to go home again and speak from experience. But in our case, for the first time since 1917, teams of foreign technicians were brought in to live and work side by side with Soviet citizens — not temporarily but for years, and not in some disembodied administrative office but where the work was actually done. There was considerable risk that in time the foreigners' "impressions" would give rise to afterthoughts, so the nervousness of the Soviet oil people was not without foundation.

Conrad visited the U.S.S.R. three times, in 1931, 1933, and 1936. The welcome he received on the first two trips was genuinely triumphal. From the academician Gubkin to the geologists on duty, all those who counted in Soviet petroleum outdid each other in overblown panegyrics. Certain oratorical traditions contributed to this, but the spectacular and almost immediate success of electrical coring had much to do with it.

It must be said that in putting their money on a foreign (and, more importantly, a practically untested) technology, those in charge of the various oil trusts had really staked their all. The produce-or-perish equation did not leave them much choice. In the circumstances then prevailing in the Soviet Union, where the abnormal requirements of the Five-Year Plan sowed confusion and the disorderly procedures of the *apparatchiki* (bureaucrats) jammed production (the G.P.U.'s knout, meanwhile, stimulating zeal at the job), there can be no doubt that the efficiency and high yield of the "Professor's" method were a veritable windfall for those concerned. There was nothing feigned about the deluge of praise to which Conrad was subjected. And what could not fail to add to it was his genuine sympathy for a social experiment that struck him as innovative in every respect — a sympathetic interest to which his counterparts quite naturally responded.

Conrad voiced this attitude on many occasions, not only in Russia but in France as well. He was amused by the long faces and annoyed looks of certain persons whom the tenor of his thought disconcerted, and I know of some in Paris who wondered whether the homage he received in Moscow might not explain what they called his "infatuation" with the Soviets. Had he not contributed to the circulation in France of a Soviet film on juvenile deliquency — *Le Chemin de la Vie* — which, by the way, was a classic in its genre? But, besides the fact that Conrad did not have it in him to repay those who flattered him by praising their ideology, he was both too clearheaded and too modest to pose as a political theorist. What interested him far more than the nature of the Soviet state was the ubiquitous effort to put the regime's economy on a solid foundation. Nevertheless, while he brought to his work in Russia an integrity without ulterior motives, and while he used the full weight of his prestige to induce more than one reluctant engineer to get on with his work, it does seem that (at least with regard to the Soviet Union) he confused industrialization with socialism — besides the fact that from the Caucasus to Kamchatka, a unique field for experimentation was offered to *Pros.*

Conrad's interest in the Revolution had not waited for the signing of contracts with the oil trusts in order to manifest itself, but this event certainly helped. Those contracts, along with research and tests over ideally diverse terrains, made it possible for *Pros* to survive during the lean years. For this Conrad felt a discreet gratitude, whence perhaps came his tendency to idealize the Russians as a people. Even so, he was not the last to recognize the difficulties inherent in a regime that disappointed him. Although he did not talk much about this it could be felt, or divined, because his enthusiasm had noticeably diminished on his third journey. But even earlier, when collaboration with the Soviets was at its peak, Conrad had doubted that the arrangement could continue without snags. Maurice Martin, a veteran of the experience in

the U.S.S.R., assures me that as early as late 1931 Conrad saw that the
development of *Pros* would depend on a tremendous push in the direction of the United States.

From Grozny, where the first team of five engineers was installed,
activities were soon extended to the regions of Baku and Emba, then
from there to the Donets Basin, the Urals, and Central Asia. In 1930
there were eleven engineers in place, in 1931 fifteen, out of the total of
twenty-four then employed by *Pros*. A considerable quantity of equipment had been sent from France. Surrounded by a good many Russian
co-workers, the crews prospected the terrain, cored the boreholes,
instructed the Russian technicians, and prepared for the local manufacture of equipment. Before long, at Baku, spontaneous polarization was
tested, and experiments connected with pressure and temperature in
the wells (teleclinometry and lateral coring) were carried out. An out-
line of the program, submitted in 1933 by the N.G.R.I. *(Neftianoe Gul-
ogorazviedochnyĭ Institut)* in preparation for Conrad's second trip to
the U.S.S.R., shows the range and development of the works planned
or under way. The whole arsenal of geophysics seemed destined to
become involved, from the study of salt domes to the depth of perpetual
ice. Around the middle of 1935, the number of electrical corings performed in some 1,800,000 meters of drilling exceeded 7,000 meters; surface prospecting had covered more than 19,000 square miles. At the
same time, construction of equipment in the U.S.S.R. had increased so
much that most of what was being sent from France consisted only of
the most recent prototypes. In addition, the training and experience of
the Soviet cadres improved continually, so that the role of *Pros* personnel (with about forty crews scattered throughout the Soviet Union) was
reduced to that of consulting engineers.

In January 1936, shortly before going to Moscow to negotiate the
fourth and final contract with the Commissariat of the People for
Heavy Industry, Conrad took pleasure in setting down, in a memorandum written for his own use, some of the findings he retained from
this really remarkable collaboration. He noted that as early as 1929, the
Russians had perceived the importance of electrical coring for the petroleum industry, whereas it took the Americans five more years to be
convinced—a change of mind brought about, in part at least, by the
results achieved in the Soviet Union. He described the cordiality of the
people he met, their natural kindliness, the open-mindedness of young
men strongly attracted to science and technology—a climate, in brief,
in which the spontaneity of the persons dealt with and the ease of relationships did much to charm him. What strikes me in this memo is

the fact that he apparently wanted to credit the happy issue of a long
and fruitful association to the Russian people alone. If, on the technical
level, the palm belonged to *Pros* personnel, on the human plane it
belonged to those they had worked with in the U.S.S.R.; and it would
seem that by enlarging upon their good qualities he wished to contrast
them with the "terror" of the police state.

If I read the signs correctly, Conrad's predilections were not shared
by the engineers. They went to the Soviet Union as they would have
gone to Tierra del Fuego; they had a job to do, and they did it as well
as they could. They were closer to day-to-day reality than was Conrad,
and not much inclined to intellectual speculation: Their view of things
was less abstract than his. They were young men with no experience
outside of their native land, thrown overnight into an impenetrable
(because different) world where they had everything to learn, including their job; they had neither the intellectual background nor the lei-
sure to think out their ideas. The harsh living and working conditions
were also factors. In the end, even those whom an initial curiosity had
induced to choose the U.S.S.R. over some other country now made little
or no effort to understand the people. Everything contributed to this
indifference: the language barrier, the ideological conflicts, and their
own lack of political sophistication as well. In this regard it is interesting to note that their literary eclecticism — Valéry and Ruskin, Stendhal
and Byron, Spanish grammar and piscicultural hydrobiology — did not
include a single work on the history of Russia; since they ordered their
books through *Pros*, I would have been aware of it. As for magazines,
they subscribed to *Candide, Gringoire, Je suis partout, Match,* and the
like. If there were any exceptions I am not aware of them. While their
letters and stories were about customs and people, they did not go
beyond things they saw, and made no attempt to analyze facts within
an overall perspective.

In those years, the first contact with the world of the Soviets was
anything but inviting. Most of the engineers mentioned a vague feeling
of malaise as they crossed the Polish-Russian frontier. No detail of landscape distinguished one country from the other; the same plain and the
same birch forests stretched out on either side of an invisible line — and
yet, as if one had passed a point of no return, the air and light seemed
to take on an oppressive quality. It could not be the slogan "Workers
of the World, Unite!" on a banner straddling the railroad tracks. It was
the finicky verification of passports, the laborious scrutiny of stamps,
the meticulous inspection of baggage, the suspicion aroused by the least
bit of printed matter, the forbidding expressions of the guards in their

ankle-length capes, the rifles, the bayonets, the watchtowers, the transfers from one train to another, the blinds drawn in the car windows once the frontier was crossed . . . pettifoggery, gratuitous vexation built into a system? No, not even that. There was nothing personal here, nothing directed at any particular traveler; rather it was a kind of morose aversion to anyone coming from outside.

Arrival by sea was equally lugubrious. Martin tells me about the slight feeling of distress he experienced when, as the only passenger on board after the stop at Istanbul, he found himself face to face with a line of militiamen when he disembarked at Batum — a ramshackle, almost deserted town without a single vessel at the quayside. He was armed with import licences; a huge fellow in an enormous cap had come from Grozny for the sole purpose of expediting the procedure, but to no avail. It took two days of parleys to clear his prospecting equipment through customs.

At any rate, on the way into Russia, the apparatus and the indispensable technical documents did finally overcome the inertia of administrative formalities; on the way out, however, leaving the U.S.S.R. for France, the slow pace of the exchanges was beyond understanding. At times, it took a telegram ten days, a letter three to five weeks, and a package several months, to reach the Rue Saint-Dominique. So capricious were the obstacles that it fell to Paul Charrin, who was in the second group of engineers to arrive in the Soviet Union in April 1930, to shuttle between the two countries for years in order to maintain a workable business relationship.

Once inside Russia, Martin lost no time getting to the place where he was to work. In the Caucasus there were two crews whose task it was to study the plain of the Terek, north of Grozny. Gravimetric prospecting had revealed nothing in this area, which was without significant outcrops as far as the steppes of Astrakhan. Each crew had two Ford pickup trucks at its disposal. The labor force was recruited locally in the Cossack villages where the prospectors had established their camp. From the month of May on, after a short, luxuriant blossoming, the heat dried up the plain; it then became a flat, greyish-yellow immensity, dotted here and there with low hillocks from which, in former times, the Cossacks had kept watch. But there were also sand dunes cutting across the plain. Extending the cable in a straight line over these dunes was something like a tumbling act. Starting off at full speed, the vehicle bounded to the top of the dune, careered down the other side, picked up speed to climb the next dune, then repeated the act, unrolling the cable as it went. If the truck skidded or got bogged down in the sand, the whole crew put their weight to the wheels. Such accidents threw the equipment out of working order, so that all too often the cable had to be hauled by hand over hundreds and hundreds of yards, in scorching heat. A sandwich and a gulp of tepid water held body and soul together. The crews set out at dawn and came back at nightfall. The column of vehicles raised a plume of dust which the setting sun turned to rainbow colors. While the Frenchmen succumbed to fatigue, the Cossacks had enough energy left to dance by torchlight far into the night with the local girls, to the music of a balalaika or an accordion.

The configuration of the land and inadequate equipment some-times made an exhaustive exploration impossible. For instance, cables could not be thrown across the Terek because of lack of boats and suit-able moorings. On both sides of the river, strips of ground several miles wide had to be left unexplored; it was only years later that the Soviets discovered a geologic anomaly likely to conceal a productive deposit. Moreover, as crews went farther and farther from their base of opera-tions, working conditions, lodging, and food supply grew more precari-ous. The villages were farther apart; provisioning was uncertain and ruinously expensive. An egg cost one ruble uncooked, two rubles cooked; a pound of tomatoes went for five rubles; there was little or no meat, and canned goods for basic nourishment were hard to find. Our engineers figured their monthly expenditures at about 500 rubles— (*Pros* allowed them 350. As for the Soviet workmen, who were paid by the oil trusts, their wages had a ceiling of 70 rubles a month, but they were expected to obtain what they needed in the "free" market.

The work camps were strung out one after the other, no two alike. After the Cossacks' small houses, which were whitewashed outside and clean inside, those of the Chechen looked dirty and badly kept up. No dancing here, no night life. Now and then, gunshots punctuated the darkness: vendettas, habitual with the Chechen, but banditry as well, often directed against the Soviets then. Farther north, it was the terri-tory of the Nogai nomads, who wandered with their herds over the steppe of the same name. They could be encountered at Makhmud Mekteb, a water source where the prospectors had set up their tents. The water there smelled of rotten eggs and was undrinkable; when it was boiled, it lost just enough of its sulfurous odor to make a not-very-tasty tea. The tents were of Soviet manufacture, the camping gear French; hurricanes and torrential rains sometimes damaged them. Typhoid fever and malaria also threatened the men.

Like Conrad, Marcel abstained from pushing into these remote areas on the three trips he made to the Soviet Union between 1929 and 1933; insurmountable difficulties were the great distances, the wide dispersion of the crews, the difficulty of getting into the areas they were prospecting, the deplorable hygiene and provisioning.

So it was up to the engineers to meet Conrad or Marcel in Moscow or Grozny, Tiflis or Baku. They came in from all points of the compass, often after exhausting runs to catch a train that was always crammed full. For example, there were Jacques Castel and his driver, who, on their way from Sterlitamak to Ufa in the Urals, almost froze to death in their sleigh, muffled up though they were in *shouby* and other furs that smelled strongly of billy goat. There were no signposts to mark the road; it was only by trusting their horses' instinct that they came through alive.

There was Georges Guichardot, who, leaving Aktiubinsk in Kazakstan with his truck for Dossar and Guriev on the Caspian Sea, was brought to a halt by a frozen ball bearing. So, he proceeded on foot to a *sovkhoz*; there was interminable haggling to obtain a horse and cart, then twenty miles to Novoalekseyevka, with the prospect of holing up there all winter for lack of transport. By the purest chance, two heavy trucks made a stop there, and they had room for him. Twenty hours of jolts and bounces, breakdown after breakdown, punctures, broken springs, oil freezing, a near-fall through the ice crossing the Uil, one of the trucks out of commission. Guichardot, who was not long on patience, finally conceived a dislike for the Russian drivers. "In an open space almost as big as the Place de la Concorde, but empty," he wrote to Charrin, "my driver found a way to be run into by one of his colleagues: front mudguard torn off, wheels in splinters, axle bent."

There was Maurice Martin, at the time an engineering consultant to a Soviet crew exploring the valley of Ferghana in Uzbekistan, who made his way to Baku by rail and ship—1,400 miles via Kokand, Leninabad, Samarkand, Bokhara, and Ashkhabad, embarking at Krasnovodsk for the crossing of the Caspian. The railroad equipment and tracks had been almost entirely destroyed during the Bolshevik Civil War (1918–1921) and could not meet the demand. Even when an official agency procured tickets for you, finding a seat on the train required complicated maneuvers and days of waiting in the stations. There were sleeping cars of the Orient Express type; soft cars with four berths; hard cars with swinging doors; and—lastly—fourth-class cars, boxcars, for animals and men together. What with changes of train and halts in open country, the chances of winding up on a siding were frequent. Thirty kilometers an hour was a respectable speed. On the train, life was organized around a stove serviced by the *provodnik,* a personage who functioned as conductor and general factotum. Each passenger unpacked his food, uncorked his flask of vodka, ran to fill his teapot at the tank of boiling water in the station, and tied down his valises and bundles to discourage thieves. Every place was alive with turbans, caftans, black veils, and vermin — picturesque but sad, a poverty-stricken Orient. And it was no different aboard the moth-eaten vessels that plied the Volga and the Caspian, crowded to bursting, as if all of holy Russia had been bitten by the travel bug.

In the long run, inevitably, living and working conditions took their toll on the morale and health of the prospectors. In some cases illness, in others overwork, forced a rotation of personnel. But it was not easy to deal with this, in view of the fact that after May–June 1932 the Soviet authorities began to refuse visas to some of the engineers appointed to replace those who were leaving. So, without explanation or apparent cause, Guyod, Bayle, Puzin, and still others were denied entry into the U.S.S.R. A maneuver to hasten the formation of crews made up entirely of Soviet citizens? Perhaps. What did it matter that the cadres were inadequate, the drilling equipment in lamentable condition, or tools and special steels scarce? Once the planners had fixed oil production at x tons, it was that or people's heads. The geologist who put a casing down an unproductive well risked losing his life. Since everything that had to do with electrical measurements came under the authority of the French engineers, while responsibility for decisions made in the field fell to the Soviet expedition heads, it is obvious that a great deal of tact and judgment were demanded of the former in order to prevent the slightest hitch that might be charged to the latter. I think our people succeeded in this to a very large extent, because so far as I know, not one of them was ever called to task. One day at Irkutsk, (with the temperature 47°C below), a Russian technician teased Jost, "Do you know who gave us the idea of looking for petroleum at Lake Baikal? It was Jules Verne." This anecdote seems to me to point out the good relations that had come to exist between the Soviet workers and the Schlumberger engineers.

At the other end of the scale, nothing appeared to hinder business with the oil trusts. After Groznyeft, contracts were signed with Soyuzneft, the N.G.R.I., Glavneft. Melikian coordinated, Charrin shuttled, Mathieu drew up reports and more reports ("with the Russians, the more the merrier," he wrote). Conrad and Marcel multiplied their lectures and expositions before a gathering of geophysicists, the engineers measured and instructed. Indeed, everything seemed cordial, so much so that here and there extraprofessional contacts were made (indirectly, it is true, with Melikian as an intermediary). For instance, some office would ask *Pros* to find certain technical books or to pave the way for an exchange of paleontological collections with the Musée d'Histoire Naturelle in Paris.

The climate, however, was changing faster than was apparent. Suddenly, in 1934, there were only four French engineers in the Soviet Union, and in December 1936, only three. These figures show better than any commentary that the experiments were drawing to a close. All payments coming from the U.S.S.R. ceased in May 1937. In September, Melikian's apartment was put under seal by the customs administration, which demanded 34,000 rubles from him as payment of alleged taxes due on some personal packages (these, by the way, were exempt from customs duties by the terms of agreement in force). The very enormity of the sum demanded (equivalent to the salary of a Soviet worker for a third of a century), which *Pros* immediately guaranteed, smacked of provocation. Registered letters tracing the history of relations between *Pros* and the oil trusts were addressed to the Commissariats of the People for Heavy Industry and Foreign Commerce, to Customs, to the Bureau of Geophysical Research, to the Soviet Embassy in Paris: No answer.

Fearing for Melikian's safety, Paris, on October 10, 1937, filed a visa application for Léon Migaux, director general of *Pros*, hoping that his presence in Moscow would smooth things over. Maurice Martin, who knew the Soviet dossier and spoke Russian, was to go with him. Migaux's visa was held up for several months, Martin's was refused. One year later to the day, *Pros* was informed that since April 29, 1936 (the date of the last contract signed by Conrad in Moscow with the oil trusts), *Pros* had caused the trusts "a material loss of approximately 400,000 rubles." "We are not giving you a detailed account of this sum," the letter from Glavneft added, "nor of the illegal transactions of your agent Melikian." Eight months earlier, Vahe Melikian had disappeared without trace.

11

Electrical "Logging": From Maracaibo to Burma

PARADOXICALLY, electrical coring—now called "logging"— which was to launch the rapid rise of the Schlumberger business and to become its principal activity, had at the outset been thought of simply as a system complementing surface measurements. What Conrad saw in it was a technique probably worth further study but essentially secondary; mine prospecting had been improved but was not about to be transformed by it. Although the Péchelbronn experiment in September 1927 had suggested that a method whose practical applications were still uncertain had been discovered, no one came within miles of foreseeing its revolutionizing effect on the oil-producing industry.

Electrical logging was beset with a host of initial difficulties including rudimentary equipment, the character of the terrain where the method was tested, the almost unanimous reservations of geologists, and the time of its introduction—shortly before the 1929 Crash. Of the four attempts to put it into practice during that year—in Venezuela, California, Oklahoma, and the Dutch East Indies—only the first (initially undertaken for Shell Oil) gave encouraging results. Venezuela, where extraction of crude had risen in ten years (1918–1928) from 320,000 to 100 million barrels annually, was then the second largest producer of petroleum in the world.

Deschâtre and Bayle—the same Bayle who, as I have noted, had "won" Venezuela by tossing a coin with Sauvage—were four weeks en route from Cherbourg to Maracaibo. Since this city, second in population to Caracas, had no hotel worthy of the name, they were lodged in one of the comfortable bungalows built by Shell for its personnel in

Haticos, the "chic" suburb of Maracaibo. Three days later they were at Cabimas, on the western shore of the lake, with their cases of equipment. The best way to get there was by flat-bottomed barques that danced disagreeably on the waves. Shell had also established a camp at Cabimas — bungalows built of rough timber but well appointed. The local work force lived in Cabimas itself, a village built on pilings like the ones that Alonso de Ojeda and Amerigo Vespucci saw there in 1499. The engineers and the geologists seldom set foot there. At Cabimas and La Rosa, a field close by, two blowouts had polluted hundreds of acres of the lake, destroying the fishing upon which the people around the shore depended for a living. Now, most of them worked for the three principal oil companies exploiting the area: Shell, Gulf, and Standard of Indiana.

At Cabimas, the *Pros* engineers' only concern was the problem of correlations. This was fortunate because they were not advanced enough to make more sensitive studies; the wide variety of information that electrical logging would furnish was still in the offing. On the other hand, little as the terrain might lend itself to the task, our prospectors were reasonably sure that their measurements of resistivity would reveal the limits of the formations traversed, thus making it possible to establish markers for correlations between drillings. Now it happened that the geology of the terrain and the twenty or more wells in operation at Cabimas offered Deschâtre and Bayle a field of exploration much to their liking. Hoping to produce curves that would show the Shell geologists how far they were ahead of them, they rolled up their sleeves and went to work.

That first generation of Schlumberger prospectors certainly possessed the pioneer spirit. Fresh out of school as they were, I don't know what had prepared them for it. The living and working conditions of the time were nothing like those of today. Climate, lodging, food, solitude, equipment problems — these discomforts and obstacles were their daily lot. Being at once engineers and handymen, jacks-of-all-trades living on quinine, they had to be able to do everything with the means at their disposal. Hand-operated winches, breakage of cables, point-by-point measurements, transfer of curves to strips of graph paper, breakdowns of all kinds. . . . There was little indeed to make their work easier. Ten, even twenty hours on the job without a break was nothing unusual. I often wondered at the small number of desertions among these men, to whom an upset stomach or a toothache was something they could not allow. As an assiduous reader of their reports, I admired the fact that, underneath the gnashing of teeth that came through, the assurance of success always made itself heard. It was as though, being scattered to the four corners of the earth, they had made a pact with themselves to hold out against wind and tide. Here, for instance, were Deschâtre and Bayle, cabling Paris — no code! — on March 5, 1929! "Success, Hip Hip Hurrah!" Electrical logging, as they had just succeeded in demonstrating, was an irreplaceable instrument for correlation. Some years later, at the La Rosa field, a commemorative plaque would mark the scene of the "world premiere" of logging.

Nothing succeeds like success. One success bringing on another, Gulf and Standard of Indiana got into the competition. Significant results were obtained. The geological markers and the correlations made it possible to pinpoint the technical structures, and even, in some favorable cases, to identify oil-bearing sands. Right away, Deschâtre was dispatched to the United States to join Roche and Gallois in California, where their efforts (as I have said) would come to naught.

In Venezuela, however, where toward the end of 1939 three crews had some 800 operations to their credit, the effects of the economic crash sharply curtailed exploration and even brought it to a halt. Jean Mathieu, who went to Venezuela to take Elie Paulin's place, called the situation a "stampede." This was certainly true in the case of Shell, which called home most of its geologists and technicians. Gulf, more circumspect, was slower in cutting back expenses, and made some first-rate discoveries in the eastern section of the country. At any rate, Mathieu was three days at sea when Bayle cabled Paris to cancel his departure — but too late. Things were going so badly that Mathieu, arriving in Venezuela, found not only Paulin but Bayle and Piotte (the third *Pros* engineer) about to return to France. Left alone, he had nothing to do for three months; he took advantage of the opportunity to go over his equipment and patch up cables by the mile.

In fact, Mathieu was left to himself for a whole year, because Bayle did not come back until May 1932. At that time, most of the oil fields were on the eastern shore of Lake Maracaibo. Besides Shell, Gulf, and Standard Oil, there were some less important companies exploring — or exploiting — small fields scattered farther out in the jungle. It was Pure Oil, one of these lesser companies, that rescued Mathieu from his forced idleness. To reach the Pure Oil well, he would have to ford a stream. But even before they could reach the stream, for two days he and his assistant, a Venezuelan named Briceno, scrambled to extricate their truck, which kept falling into one mudhole after another. Having had enough of this exercise, and thinking the Pure Oil people might have a tractor that would pull the truck through, he went ahead on foot with Briceno — who had a mortal dread of the jungle — at his heels. Meanwhile, the drillers, seeing no one coming, had cemented the tubing. And, anyway, tractor or no tractor, floods had made the stream impassable.

Early in 1932, research and production began to pick up slowly. A couple of geologists of marked ability (one of them was George Murray, who came back to Schlumberger in 1936), realizing the importance of electrical logging, took an active part in getting it started in Venezuela. The hand recorder and the first offshore measurements also date from this period. Bayle, back from France, rigged up a winch driven from a truck's axle. The technology of spontaneous potential curves, introduced in Venezuela about the same time as in the Soviet Union and Romania, brought excellent results in "work-overs" — wells that are redrilled because they are poor producers.

One day, when Bayle was working far out in the hinterland for a small British company, someone came and offered him a job in Trinidad, where lenticular sands made oil production particularly troublesome. There, the spontaneous potential approach proved to be remarkably effective from the start: The curves were so clear that the very first log showed the possibility of locating productive strata. Roger Piotte came to work with Bayle, and later on Louis Bordat, who had put in four years in Russia. As the need arose, one or the other shuttled between Trinidad and Venezuela.

The work was less arduous in the island than on the mainland. Almost all the oil fields were in the vicinity of San Fernando, where the prospectors had set up their quarters. The climate and the luxuriant vegetation — "something out of *Robinson Crusoe* or *Paul et Virginie*", in Bordat's description — were in strong contrast to the West Venezuelan environment. There were the sea, the beaches, the golf clubs, the pubs — in a word, the amenities (colonial style) with which the English always managed to surround themselves. A Creole "aristocracy" with British, French, and Castilian roots lorded it over a population that was two-thirds black and one-third Hindu. The level of existence was a little less poverty-stricken, and morals and customs a little more civilized than in Venezuela, which was at the mercy of one Gomez's barbaric extortions. (When Gomez died in 1935, a typical South American revolution broke out there, after which all went on as before, with another general in power.)

There were eight Schlumberger engineers in Venezuela and five in Trinidad. Diversification of techniques and tooling led to an incipient specialization: Mennecier became an "expert" in casing perforation,

Beaufort in dipmetering. Admittedly, working conditions were no longer quite as adventurous as in earlier years, but the often primitive accommodations in the camps, the trips from one field to another by roads unworthy of the name, the equipment in constant need of care, the cables that got lost and weights that got wedged — all the toil put into producing a couple of revealing curves — tested the hardiest to the limit. Some were never able to get used to the myriad insects, the scorpions, the leeches; others, in the end, paid no attention to them. Charles Doh, who spent a year and a half in eastern Venezuela, wrote to me:

> *I remember traveling through the jungle, sometimes a week in a flat-bottomed boat to reach an exploratory drilling site, swapping canned goods with the Indians for eggs, weight for weight. I always wanted to dive into the black waters of the numerous streams that form the Orinoco delta, but orders were formal: Snakes, crocodiles and piranhas especially made swimming not recommended. One day, the natives caught a water reptile five meters long and distended in midsection by a large ventral sac; they took out an alligator longer than I am (I am 5'9" tall) and in an excellent state of preservation. During the rainy season, the shortest trip was an expedition. Once, to cross the Tonoro, I had to recruit the manpower of a whole village to carry the equipment piece by piece through hip-deep water, as well as getting the truck hauled from one bank to the other by 75 naked Indians. Imagine the tropical rain drumming day and night, week after week on the corrugated-iron roof of our shelter. No surprise if some of our people who took easily to drink fell back into the state of nature. There were fights, some were wounded, some even killed. As for me, if I escaped the long jungle knife—and that by a hair—it was because I kept a huge monkey wrench handy . . .*

Charles Doh may have exaggerated, but not too much at that. After all, a snapshot shows him, grinning from ear to ear, with a boa constrictor coiled around his arms and torso — and sure enough, the boa is bigger than he is.

While I knew something about North and South America and Russia and Romania, it was different where the Far East was concerned. Therefore, it is not easy for me to describe the prospector's life in Asia on the basis of sparse correspondence. They were blue dots on my map and I had to imagine them buried in the jungle, living an adventure even more unexpected than that of the others. So what I am going to say about them is distilled from their accounts to me long after their return to France.

When, in April 1936, Louis Bordat (en route from Trinidad) left the ship at Rangoon with his wife and children, there were already six Schlumberger engineers in Southeast Asia. The climate, the vermin, and the poverty lost nothing by comparison with what he had known in the Venezuelan brush, but living and working conditions were incomparably less severe.

Like Kuwait Oil and Anglo-Iranian, the Burma Oil Company was a state within a state. With its all-embracing organization, its agents scattered throughout the territory, no detail escaped its vigilance. As soon as they arrived, the Bordats were taken in charge with all the marks of respect due to sahibs away from home. Provided with a "boy," impeccable bed linens, flatware, and a tea service, they were then sent on by rail to central Burma. Transferring at Prome, they sailed slowly up the Irrawaddy on a paddle-wheel steamer with its teeming crowd of Burmese, Indians, Karens, and Chinese doing their cooking amid a profusion of odors that offended the nostrils of the passengers above on the upper deck. Here and there, a shining white pagoda signaled the presence, on one bank or the other, of a village buried in the tropical forest.

The oil fields of Yenangyaung (in the district of Magwe, the petroleum center of Burma) were known to the native population as far back as 1825. Yenangyaung presents an eroded terrain with no overburden, so that its geological map can be read, so to speak, with the naked eye. There is no need of complicated geophysics to delineate the wells. The Burmese, who continue to dig as they have done for centuries, position their holes with a sure hand. A reel, a rope, a pail, and a few short-handled tools, Bordat tells me, constitute all their equipment. Production by pumping out these old wells not being satisfactory, Burma Oil began to deepen them. Here, electrical measurements had to be repeated every time a well was tubed; a frightful amount of wear on the cables ensued. The infiltration of oil in the holes made the work anything but pleasant. Other explorations—in this case by guided drilling—extended under the bed of the Irrawaddy. As the cable was being hauled up, it dug itself into the wall of the hole, threatening catastrophe. On the basis of incomplete measurements, the correlations were extrapolated as precisely as possible, but this led to serious errors in interpretation.

The drillers (American for the most part) were inveterate hunters. The surrounding jungle teemed with tigers, panthers, elephants, and wild buffalo, but the men were content to shoot small game. If you were a "good gun" you gained their friendship immediately. For a couple of rupees, a swarm of children served them as beaters. The English community, a very close-knit group, preferred tennis, riding, and polo; the drinking was hard and steady.

After a while, the French were transferred to Nyangla, the "noble" quarter of the concession. Teakwood bungalows were built for them after consultation about the layout of the rooms. Bordat even won a swimming pool for his children. To me, he expressed esteem for the Chinese craftsmen; they had a strange technique and tools with shapes unknown in Europe, which they handled with great skill. Burmese and Pathans, Karens and Indians took over domestic service. Every white person enjoyed a high style of living: Bordat's household staff consisted of a dozen servants. (That, by the way, did not keep him from contracting malaria.)

The engineer who headed the missions in the Far East was Raymond Sauvage. His first assignment was in September 1933; in Digboi, northern Assam. The technician already in place in those days looked upon such "specialists," sent out by the office in London, the Hague, and elsewhere, as troublemakers. What resulted in Sauvage's case was an undercurrent of resistance — a sort of morose lack of interest in the "intruder's" methods — which did not make things easy. Of course, nothing would break the ice like a series of eloquent curves, but they had yet to materialize.

It can happen, however, that some extra-professional interest or psychological affinity — or some chance incident — may clear the atmosphere. It was Raymond Sauvage's horticultural talents (modest, by the way) that did the trick. Rosebush cuttings brought from England by the wife of the chief geologist of Assam Oil simply would not take root; they did their best to anchor themselves in the soil — as humid as any in the world — but withered in no time. Thereupon Sauvage, who had discovered wild rosebushes flourishing in the jungle, suggested that the British shoot be grafted onto its tropical cousin. Lovely roses were the result, and this won the Frenchman the favor of the English colony. After all, if he could bring off such a victory, this engineer who had dropped from the skies with his battery of bizarre apparatus might well know something about prospecting for petroleum. Thus Sauvage, promoted to gardener emeritus, soon had his hour of glory as an expert in logging.

He hardly had time to enjoy his renown: Four months later, leaving his equipment at Digboi, Sauvage was in Burma, preceding Bordat in Yenangyaung by a couple of years. Thereafter, he was constantly on the move in Southeast Asia until his repatriation in 1946. One example will suffice to give an idea of what it was like to go from one oil field to another under the travel conditions existing at that time. Shortly after his arrival in Yenangyaung, Sauvage was called back to Digboi for an urgent logging. Some 465 miles as the crow flies separates the two places, but in fact the lack of transportation forced him to make a detour to the south that quadrupled the distance. The journey, by road, rail, airplane, and ferry, took three days and four nights if all went well. One senses a tinge of exasperation in the accounting he gave of the trip: Truck to the station at Pyinmana, four hours; train Pyinmana–Rangoon, twelve hours; four-place single-motor plane Rangoon–Calcutta, seven hours; hotel, eight hours; train Calcutta–Parbatipur–Ganhati, twenty hours; ferry across the Brahmaputra, two hours; train Ganhati–Tinsukia, twenty-six hours; Digboi at last, and straight to the borehole, two hours. Eleven hours later, the job done, Sauvage was on his way back to Yenangyaung. Two weeks thereafter, same scenario over again. "You had to be in good shape," he commented — the more so since he was the only *Pros* engineer in Burma, where his work left him little time to relax.

So, in the end, he called for help, and Bégin was dispatched to Digboi. At last Sauvage, comfortably installed in Yenangyaung, thought he would have some peace; but it was not to be. In April 1934, a telegram from Paris put him on the road to Sumatra, where Jabiol had preceded him briefly in 1929. So off Sauvage went via Singapore and Batavia (Jakarta), to take up quarters at Pladiu — a big Shell center near Palembang, which included private houses, tennis courts, a club, and shops where everything could be found. He stayed there eight years, until the Japanese invasion, on the go most of the time. Given the widely dispersed work sites, the irregular coastline, and the impassable rivers, moving from one place to another was an exercise in acrobatics. It took a week to reach the northern part of the island, two weeks to get to Bandau (Borneo); once or twice a year, he made his way to Digboi. His English, picked up in the school of Scottish geologists and American drillers, is still to this day a mixture of various accents; on the other hand, being incapable of grasping the pidgin English of his assistants, he set himself the jolly task of learning Malay. By dint of seeing Sauvage race from one jungle to another, his comrades took to calling him the Mohican of the Dutch East Indies.

On the eve of World War II, there were about fifteen engineers on mission in Malaysia, Burma, Bengal, and Peshawar. None of them could get back to France before the German invasion. Cut off from all contact with Paris, they continued at their respective posts, struggling with growing difficulties caused by the progressive deterioration of equipment. With the fall of the Philippines and the Japanese landing in Sumatra in February 1942, their situation in the Indonesian archipelago became untenable. After destroying their equipment, they met in Java — just missing a freighter that had sailed for Australia. Not all of these men were stranded, however: Stoll, Leleu, and Silvestre managed to reach Trinidad after a five-month sea tour via Bombay, Cape Horn, Buenos Aires, and Rio. The others, trapped, took refuge in Bandung, where the Japanese finally put them in an internment camp. They were not too badly treated there — not as badly, at any rate, as their Dutch fellow-internees, who had the misfortune to belong to a nation that continued to fight. At the end of a year, following who knows what bargaining, the Japanese evacuated them to Indochina, where they recovered their freedom after all sorts of police harassment. Mobilized on the spot, they were assigned to different services — Palustron to the Saigon tramways, de Geffrier to the mines of Tonkin, Sauvage more or less all over South Vietnam.

The Japanese occupied the military bases along the coast and let the country as a whole sink into chaos. Most freighters and coastal vessels were damaged or sent to the bottom, so there was not enough shipping power. Although Tonkin was bursting with coal, the electric power station in Saigon burned rice and corn, which did no good to the grates. A piece of cast iron, a ream of paper, a ball-bearing gear — no matter what it was, it came from the black market and eventually went back there. Sauvage, who had charge of the inventory and distribution of materials, did all he could but could not do much; I imagine him fuming and fulminating to no avail. Even before this, his letters and reports to *Pros* had been full of sound and fury. Boris Schneersohn, the kindest man and one who holds Sauvage in high esteem, reminds me that his missives belonged in the category of epistolary terrorism. "There was never a friendly one," he tells me. "We were all mentally retarded. The equipment we produced was worthless. Those who conceived it should have given it more thought; those who produced it didn't know their business; as for using it on the ground, it left the prospector's intelligence and goodwill in tatters. All rantings and scoldings — and of the first magnitude — in a style that reminded one of Céline."

De Geffrier found in these same letters a scandalous lack of dignity. The day he too was called upon to set out for the Far East, he confided to Schneersohn that he, De Geffrier, knowing all about the problems of technology and manufacturing, would not be the one to write reports in which humor vied with invective. And in fact, his first two letters were exemplary; no one could have been more understanding or good-natured. They were, however, the only two. Those that came after were quite up to — or down to — the level of Sauvage's poison pen. One must believe that oil prospecting in the tropics brings on stylistic turnabouts. . . . All that is left to tell is that poor De Geffrier was put to death in Tonkin for resistance activities, that Sauvage made it back to Paris in 1946 after being hospitalized in Assam for three months, and that the others got back to France any way they could.

In Burma, up to the time of the Japanese invasion, the work and the days rolled on in an unreal world. The rupture of Europe, the French debacle, even the deluge of fire and steel loosed upon the cities of England, seemed not to stir the emotions of the English community in Yenangyaung. Of course, this was only appearance, but the theater of hostilities was far away and England continued to fight. In Bordat's account of all this I perceived a touch of admiration for the "fair-play" attitude of these Britons, whose life he and Baboin and Laguilharre (his two comrades) shared. The Vichy government and the *de facto* collaboration with the Third Reich notwithstanding, the Frenchmen never had to endure the slightest hint of ostracism or meet with one disparaging remark. One can imagine the outbursts of French xenophobia if, perchance, Great Britain, besides laying down her arms, had provided a launching platform for German bombers on their way to blast the cities of France. Moreover, this British microsociety manifested no doubt whatever about the final outcome. It is true that the women and children had already been shipped off to India by plane, that preparations for dynamiting the wells were complete, and that the construction of a trail passable for vehicles, in the direction of Tamu on the Burmese-Indian border, was well under way. A number of depots for food and fuel, spaced out over 250 miles, had been torn out of the jungle. Guy Baboin broke his leg while working on this project. Laguilharre had a very hard time getting him back to Yenangyaung, where they fashioned a makeshift cast for the injured limb. By pure chance, an R.A.F. airplane landed at Magwe and they were able to evacuate Baboin at top speed, after a race against the clock over a corrugated-iron road, with Baboin suffering torture at every turn of the wheels. It was a close escape; the very next day the airfield at Magwe was bombed and the runway destroyed.

No one quite knew what was happening on the Indian side, beyond the frontier. It was said that bulldozers were gouging out a road toward Tamu, starting from Imphal and crossing the mountains of the Manipur territory. It might have been one of those rumors — logical *ad absurdum*—that arise and proliferate in times of extreme tension. However that may be, Bordat saw in the junction of the two trails the chance to save the best of his equipment. He alerted the military commands of India and Burma, pointed out that the loss of Yenangyaung would make his prospecting material invaluable, and, to make a long story short, pleaded his cause so well that he obtained an agreement in principle. He loaded two trucks chockablock, leaving nothing essential behind, then camouflaged them upriver on the trail, at Kalemyo on the right bank of the Irrawaddy. Destruction of the wells and the exodus did not start until Japanese patrols had been sighted in the vicinity of Magwe. A few parachutists could have prevented both the destruction and the exodus, because the district was sparsely defended—a tactical failure all the more incomprehensible in view of the fact that a certain Kokubu, ship captain by trade but surgeon-dentist at Yenangyaung for the occasion, turned out to be the chief of Japanese espionage in Burma. He was an excellent practitioner for all that, over-generously stuffing your cavities with gold for the price of lead. Bordat still laughed when he told me about it.

It took the convoy twenty hours to reach Tamu. On the Indian side the trail was ten miles short of completion, but mighty efforts were being made to finish construction. Everybody (with the exception of drivers of rolling stock) packed his kitbag and took off through the mountains—a hardship light by comparison with what a half-million Indians who had lost everything had to endure, and of whom thousands perished from exhaustion. Once week later, Bordat arrived in Imphal at the tail end of his convoy. His refugee card is dated April 20, 1942.

12

The "Boom" in the United States

IN July 1930, the trial contract with the Big Three American companies (Shell, Gulf, and Standard) expired and was not renewed. In California, Texas, and Oklahoma, *Pros* missions were at a standstill. In Canada, surface prospecting marked time. World prices for zinc, copper, and nickel had fallen too low to keep the mining industry afloat. Even studies for dams were reduced to a minimum; hydroelectric power was giving way to common coal. As Paris viewed the scene, America was decidedly not living up to its reputation; not only had it failed to keep its promises, but work there cost money. The bottom line for 1929 showed a deficit of $3,565.95.

Profit or loss, at the Rue Saint-Dominique they always had to figure down to the last franc. *Prosélec*, the confidential periodical that served as liaison between Paris and the prospectors, made no bones about it: The May 1930 issue suggested shortcuts by which the crews could save five minutes a day! In letter after letter Léonardon, never one to give up, expressed his disagreement. "Were such prophets of doom, such mollycoddles, ever seen before?" he protested. "What is a gap of four thousand dollars?" Conscious of having raised his voice, he lowered it: "I'd better leave off the death rattles; you must be tired of hearing them and I'm tired of rattling." But then he renewed the attack, rattling as before. Did anyone, by chance, think America was his hobby? That he hung on to it because he liked hanging on? Proving that they had indeed misunderstood him, he was willing to consider taking a cut in salary, "if this was what *ces messieurs* wanted." But he would not tell his wife, "good housekeeper that she was," who would certainly make a scene. However, *ces messieurs* did not choose to pick up the gauntlet, and Léonardon's epistolary offensive reopened with enhanced vigor. "What, then, was it all about?" he wanted to know. Wasn't he, E. G. Léonardon, graduate of Polytechnique, with degrees in law, with a New York office, secretarial staff, telephone, and vast experience in American business, "the right man in the right place" to offer his services—handling commercial paper, correspondence, approaches, inquiries, negotiations, market studies—to European companies in need of contacts? All *Pros* had to do was ask around. *Pros* had plenty of contacts—and there would be the "icing on the cake," the "injection that would keep our heart beating. . . ."

Pros suggested that Léonardon take a vacation, a way of telling him to get his head examined. "I can see it from here," he groused. "They're afraid we might look like fast-food merchants." Well, they could leave it to him to grab people by the arm: "I can be trusted to do the dirty work." "The Paris clique," as he sometimes called his correspondents, turned a deaf ear. On September 6, 1930, Léonardon was asked to ship his logging material to Venezuela and send his flock home to France. "I think," he wrote to my husband pleadingly, "that you are still close enough to your stay in the United States to tell *ces messieurs* firmly that their plan sounds like a joke." Joke or not, he could get nowhere. One month later the entire Schlumberger personnel had left the United States. Léonardon alone stayed behind to keep the flame alight.

The blue porcelain-headed pins in my wall map were spaced farther apart. Outside the U.S.S.R., mine and petroleum prospecting had slowed almost to a halt. Where it kept going at all, the time was not favorable to new techniques. At the Rue Saint-Dominique, where research had to continue whatever the cost, the perpetual question of finances again became a worry. The economic crash, which killed off a large number of businesses, forced others to emerge. One of them, the Société Géophysique des Recherches Minières (S.G.R.M.) made overtures to *Pros*. An agreement was facilitated by the fact that its activities (which involved seismic methods and magnetometry) did not encroach upon *Pros*'s domain, but in a way complemented theirs.

The negotiations, speedily conducted, led to the creation of the Compagnie Générale de Géophysique (C.G.G.). The terms of the agreement retained measurements in boreholes for *Pros* and assigned surface prospecting to C.G.G. Schlumberger brought to the new company its technology and know-how, S.G.R.M. its tools and licenses, and—most important—some 5 million Poincaré francs. This was the first sum of money that did not come directly from the family. A third company, the Société de Prospection Géophysique, which was experienced in dealing with gravimetric problems, joined the group later. As time went on C.G.G. would come to control all French surface geophysics.

Working capital now assured for the short term, it was possible to keep the machinery oiled and running even where returns were uncertain. But the end of the tunnel was by no means in sight. Most of the crews still in the field were working at a loss. The few companies that showed any interest in electrical logging slashed prices mercilessly. Taking another tack (for promotion purposes, but also because they sorely needed the opportunity to experiment where boring was going on), *Pros* proposed to interested clients that it would assume part of the cost of testing electrical logging. I do not know how this initiative was followed up. Revenue coming from the U.S.S.R. and Romania, cutting down on personnel and salaries, Conrad and Marcel's unpaid work, and the strictest possible management, together were not enough to balance the books: Operations in 1931 and 1932 showed a deficit.

Starting in the last third of 1932, however, the descending curve seemed to be leveling off. Two years of reduced production had seriously lowered world petroleum reserves. Resumption of drilling was at hand, and with it—although as yet hardly perceptible—a renewal of exploration. In Venezuela and Trinidad, the efforts of the crews were beginning to bear fruit. Surface prospecting, henceforth left to C.G.G., was catching on in Algeria, Morocco, and Gabon. After twenty months of absence, a crew was back in California; in January 1933 another crew went to work in Texas. There the broad plain along the Gulf of Mexico—the region that oilmen call the Gulf Coast, extending from the Mississippi to the Rio Grande—was to be the scene of the first spectacularly successful use of electrical logging in the United States.

The contract negotiated between Paris and Royal Dutch Shell in The Hague was not without its drawbacks. In return for a lump-sum payment and various provisions (such as the use of vehicles, additional personnel, and lodging), Shell secured not only systematic logging of all of its wells as they were drilled but also a substantial percentage on operations carried out for third parties by our engineers. This clause, which made it difficult to make a profit on work done for anyone but Shell, in effect put our crews under Shell's control.

And, as in 1930, the geologists' attitude was one of hostility. Jean Mathieu, arriving in Texas from Venezuela, was greeted with the prediction that the Schlumberger methods did not have long to live! These people were firm in their conviction that since there were few faults in the subsurface under consideration, there was no possibility of estab-

lishing correlations between wells. Mathieu was familiar with the argument; it had been the same three years earlier. The only thing new was
that this time he had the tools with which to answer it. With a one-
ton tricable, a mechanized winch, and a recorder that traced two simul-
taneous curves, he had a good chance to convice the geologists that they
were wrong. Anyway, his detractors could not do a thing: Orders came
from The Hague and it was up to them to comply. "I have a commission to fulfill," he said, "give me what I need and the time to fulfill it.
When the moment comes, if it's all right with you, we'll examine the
results."

They gave Mathieu a driver, two assistants, a Dodge patched up
for the occasion, and an old four-wheel-drive Mack truck. He installed
his equipment on the truck, hurried to New Orleans to meet his fiancée
as she landed from France, married her on February 7th, took to the
road on the 9th, and made his first measurements on the 11th. At the
Iowa field in Louisiana, some thirty miles from Lake Charles, his sixth
borehole not only indicated a salt dome but also rendered clear correlations and located faults.

Proud of his success, Mathieu cabled the news to Paris, and Paris
made haste to pass the word to The Hague. Two weeks later he was
summoned to the Shell geology department in Houston. He expected
congratulations, but instead was received rather cooly. Thereafter, he
was told, he was to hand over his diagrams as soon as recorded to an
agent of the company, at the drilling site, without keeping copies. At
first, Mathieu thought there must be a whiff of bad humor in this;
these folks had probably been taken to task for making a mistake and
were saving face by playing the high-and-mighty. In reality, however,
the move to get hold of the diagrams at the source came from the high
command at Shell. They were afraid—and the fear was soon shared by
all the big oil companies—that Schlumberger, with the abundant information gained by its prospecting methods, would be tempted to stake
out potentially productive land.

One of the factors that was eventually to guarantee Schlumberger's
reputation was that the company never departed from its limited role
as consultant, but this it had not yet had time to prove. Another of the
leading oil producers' worries—this one with more reason behind it—
was that the results of the electrical measurements made in their wells
would fall into the hands of the independent operators, who indeed
spared no effort in that direction. In March 1933, Léonardon informed
Henri Doll that Deschâtre had been approached from various sides by
scouts with fat bankrolls. As early as April 1930, *Prosélec* counseled

everyone to keep the magazine and all technical documentation out of the reach of indiscreet persons. Later, Conrad reminded all expedition leaders of the strictly confidential nature of their work:

> While inadvertent leaks might be most to be feared, prospectors must be no less aware of the danger that spies might get into their offices without their knowledge, and make copies of documents without leaving a trace of burglary. Such hidden theft, which is more to be dreaded than open stealing, leaves us without a moral leg to stand on vis-à-vis our clients.

The prospectors suffered more than their share of robbery and pillage; finally, technical literature was no longer sent to the field.*

In the United States, oil prospecting had for a long time been a hit-or-miss affair. Until 1930 nine-tenths of the country's thousands of wells were drilled without recourse to geophysics. On the other hand (as if the discovery of one reservoir must necessarily bring others to life), the minute a pay load was found the scramble for wealth and glory began. Hundreds of individuals came running, grabbing up parcels of land for miles and miles around. From one hour to the next the lease price of an acre was multiplied twenty to fifty times. Derricks proliferated. Each prospector kept his eye on neighboring rivals, counted the number of lengths of pipe they drove into the earth, calculated thereby the depth of their holes, and, if by chance a gusher came in, everyone around drilled to the same approximate depth, hoping for a stroke of good luck. Very few came out of this wild free-for-all with anything to show for their pains.

In 1935, when most of the industry no longer questioned the merits of electrical logging, the discovery of a field that was worth the trouble would cost the big, particularly "able" companies on an average of $400,000; the slightly less capable companies about $2,000,000; the independent operators $8,775,000; and the promoters who juggled with other people's money over $37.5 million.

In 1933, the cost had to be still higher. So, as the result of the favorable signs of the times, Schlumberger experienced what amounted

---

*This also happened with regard to *Cégésec*, the bulletin devoted to surface prospecting. Both bulletins, which were written in English beginning in November 1936, ceased publication after the invasion of France. In 1950, the Research Center at Ridgefield began to publish the *Schlumberger Technical Review*, an equally confidential document.

to a boom—the first in its history. In the state of Oklahoma alone, receipts went from $500 in January to $8,591 in December. The boom was a great deal more spectacular in the Gulf Coast area. Was interpretation of the curves still inexact? Unscientific? Roughly quantitative altogether? What did it matter? There were those who yesterday had drilled at random, and today posted themselves at the roadside like hitchhikers, chilled to the bone, waiting to intercept the passing trucks: "Hey, wait a minute, Schlumberger, they're looking for you!" At West Ranch, at Old Ocean, Anahuac, Beaumont, wherever electrical logging was called for, the Schlumberger engineer was taken for an oracle. Louis Allaud tells me that one of his clients brought his wife to the well. She was unable to contain her impatience and cried out, floundering in a sea of mud, "Hey, Frenchie, is there any oil? You're going to make the damned oil spout for us, right? Because, if there is any, my husband buys me a pearl necklace. . . ." She was not the only one to have her head turned by a sniff of oil: It was not long before sober bankers offered independents not pearls but substantial loans, with nothing but a promising diagram to go on. I am told that there was even a black market in these diagrams.

In any event, when a dozen or more deposits were discovered on the Gulf Coast in 1933, over forty oil companies were formed that same year in Texas alone. A lot of drilling and logging was therefore in sight. Léonardon, who had foreseen this, predicted it, sworn by it, was loud in his triumph. Ha! Now that the U.S. was forging ahead again, he wrote, "*ces messieurs*, disgusted with the U.S.S.R., are talking about invading America in forces in order to kill the competition in embryo by a massive attack." In an excess of optimism he attributed this warlike language—familiar enough in his own style of letter-writing—to his correspondents; it appeared even in his coded telegrams. Léonardon applauded wholeheartedly upon finding Paris finally beginning to see things his way. Not that he was priding himself on all this, not exactly, but after all. . . . I laughed to myself over the circumlocutions of his prose. Reading between the lines, one could see him thrusting out his chest. Who was it that had pleaded against demobilizing the crews? Refused to pack up and quit? Never stopped furbishing his arms? Well then, urging him on to the offensive was like preaching to the converted. Except that in his humble opinion—"and my opinions will become more and more humble, for in spite of the fact that I am no expert in American matters I can't make a proposal without having it answered with a counterproposal"—in his humble opinion, attacks and invasions called for troops and more troops. But how did Léonardon stand in the matter of troops? His assault forces consisted of four pro-spectors worn out with fatigue: Deschâtre, Gallois, Henquet, and Mathieu. So . . . since *ces messieurs* seemed ready to admit that they had taken the wrong turn, was it too much to hope that, having learned their lesson, they would be a little less zealous in telling him how to do his job? And that in order to lay siege to America in style, they might begin by sending him reinforcements in men and material?

The problems of recruiting and training engineers had unexpect-edly become acute. Suddenly, the same urgent appeal came in from all sides at once: "We don't have enough experienced professors." Léon-ardon, who saw his personnel go from four to seventeen in the course of a few months, was not the last to complain. He produced impressive figures and tables: 5,017 miles worked over; 340,861 feet of boreholes logged; 1,709 hours of work on the ground, and so on. "All the holes brought in by the lucky independents," he wrote "come to us. The crews are overtaxed. This irregular rhythm, which lowers the quality of the interpretations, hurts us to a degree that you cannot appreciate at a distance."

One of his complaints had to do with the recruits' lack of language competence, which sorely affected relations with clients. Gillingham is English but talks British*; Léger jabbers a bit of English but doesn't understand Texan. Before long the question "Do you know some English?" was part of every interview, and an affirmative answer helped a candidate almost as much as his diploma. Yet even to these favored ones it was charitably suggested that they refresh their vocabulary.

In less than a year, thirty-five new recruits were initiated as well as possible into the methods of electrical logging and sent out to wher-ever they were most needed. In October 1934, Paris wrote to the expe-dition leaders to say that its personnel reserves were down to zero. Thereupon Léonardon, disregarding the opposition of *ces messieurs*, hired his first American prospectors.

The opposition to this move did not come exclusively from Paris. The oilmen in Texas and Louisiana distrusted local technicians. As they put it, it was a grave mistake to trust them: They had so many contacts and connections that they could easily drop a hint here and

---

*London speech is nothing like that in Louisiana. After Gil-lingham arrived, one of the local companies besought Paris to send a French engineer who spoke English that they could understand: "Can't you send us one of your goddam Frenchmen?"

there with some benefit to themselves. As I have already noted, the minute a significant logging was made there was a strong temptation to slip the news to a pal lying in wait. Therefore, wily Léonardon thought it well to make the rounds of "rural" universities, as he called them — in Iowa, Ohio, Missouri — on the lookout for recruits who had no connections with the oil business. "None of those wise guys from the civilized Northeast," he explained to me. "No, I wanted some lads from the Midwest, farmers' sons, not smart alecs who would leave you for a song. But did I ever get hell when I took on the first of them: "That does it, we're finished, the Americans have us where they want us," wailed the Paris clique. Note that there was something comical about this, because we had just set up the Schlumberger Well Surveying Corporation, a one-hundred percent American company. And anyway, how did they expect me to get along, what with the speed at which we were growing?"

There was also another aspect to the problem. Not only was Léonardon short-handed, but he could not employ the new arrivals from France without breaking the law. There was such a demand for immigration visas, and it took so long to get them, that in order to gain time the prospectors came in as tourists and went to work under false pretenses. A Houston lawyer found an unexpected substitute for this ploy, which otherwise could have turned out badly. Not that our Frenchmen, who were outside the quota, would not have applied for admission in due and proper form; but that, because they had to get their visas from the U.S. consul in France, to whom the applications were made, they would have had to cross the Atlantic twice. Not at all, the above-mentioned lawyer decreed; it will cost you only three or four days instead of four weeks. All you have to do is hop over to Matamoros on the other side of the Rio Grande, establish a fictitious domicile, head for the American consul, ask to have your file transferred to Mexico. There's no objection, it's according to the rules, and one telegram takes care of it. Whereupon the applicant for immigration crosses the border again as a tourist, goes back to work, and waits till everything is ready and he is notified with his fake domicile. The amusing part is that when the consul general in Paris began to suspect some sort of funny business in the series of telegrams from Matamoros, my father took the trouble, documents in hand, to furnish him with some appeasing assurances concerning the role played by Schlumberger engineers in the American petroleum industry.

The year 1934 was a pivotal one in several ways. Skyrocketing business led to contract modifications with the oil companies. Doing away with fixed annual payments, Schlumberger undertook to furnish all its services on the basis of payment by unit of operation. Little by little (and with some exceptions) the drawing up of the contract, the detailed cost estimates, and even the written order, were abandoned.* It soon became customary, when a company or an independent decided to bring in Schlumberger, to mobilize a crew on the basis of a simple telephone call. Four factors determined the price of an operation: fixed charge for moving to the work site, mileage traveled, well depth, and time spent on the ground. Ordinarily, the appearance of some interesting clue in the course of a drilling operation would prompt the client's call; and since in order to do a logging it was necessary to pull the drill pipes out, thus risking a cave-in, speed was of the essence.

In the United States, where oil producers numbered in the hundreds, the procedure — except for an occasional detail — followed an almost invariable pattern. The client alerted the nearest Schlumberger center, read off his coordinates — well, depth, type of operation — and within a half-hour the crew was on its way. As soon as the site was reached the truck was backed up to the well, the various pieces of equipment put in place, the sonde attached to the end of the cable, and the operation was begun. The driver or assistant started the winch, with his hands on the brake because the drum was a freewheeling one. The engineer, protected by a huge umbrella (in later years, the truck would carry an ingeniously designed technical cabin), watched and regulated his galvanometer. The galvanometer needles were in constant motion as long as the descent proceeded without a hitch; but if the sonde ran into an obstacle — for instance, a plug of mud — and the needles stopped moving, this was a sign that the cable was in danger of becoming tangled. Recording the measurements took place as the sonde came up the length of the hole. These measurements, which nowadays are picked up on magnetic tape and transmitted directly to the computer by telephone or satellite, were at that time summarily interpreted on the spot. The Schlumberger engineer, diagrams in hand, evaluated the reading under the critical eye of the client — maybe there's water here, maybe gas there, maybe petroleum between these two points; I'm going to study that more closely. . . .

The geologist on duty was not usually too convinced; frequently he had to have proof that he could handle and taste — in a word,

---

*At the present time, after forty years, there has been little
 change; the contract, in the usual meaning of the term, inter-
 venes only in negotiations with state monopolies.

mechanical cores. A gun with hollow bullets was lowered, a series of samples taken at the desired levels of the borehole (but there I'm ahead of my story, because the first lateral coring by sampling took place in 1936). Then the crew loaded the equipment back up and headed home; the engineer went over his diagrams again, made fair copies, and settled down to deciphering them. Exhaustive interpretation of the curves supposed a thorough acquaintance with the terrain; in order to see things clearly, the engineer had to dig into numerous reports and studies. He also had to deal with the clients, write reports, keep his equipment in good shape, and master new techniques and material brought into the world by the brain trust at the Rue Saint-Dominique. In addition to all that, if he was head of the center, he had the correspondence, bookkeeping, and supervision of the workshop, stores, personnel, and so forth.

Léonardon had not exaggerated: The men were working to the absolute limits of their strength. A recruitment policy, adequate material resources, and a large degree of autonomy were urgently needed. It was not reasonable that decisions—even trifling ones—should depend on interminable exchanges of letters with Paris. Besides that, *Pros*'s status in the U.S. was unclear (if not downright irregular); as far as legal standing was concerned, all that *Pros* maintained there was a simple bank account. There were Schlumberger Electrical Prospecting, with an address at Jean Mathieu's Houston residence, and a Schlumberger Electrical Coring in California; but realistically, these were trade names, mere fronts. In fact, it was Léonardon who, with considerable power of imagination and acting on his sole responsibility, kept the Schlumberger procedures moving in North and South America.

The growing volume of business was beginning to make these jurisdictional fictions untenable; it involved too many disadvantages and too many potential difficulties. The idea of an American company was taking shape. Subjected to study on both sides of the Atlantic, it was actively discussed and then speedily brought into being. In September 1934, my father arrived in Houston to work out a charter and bylaws with legal counsel. My mother and my sister Sylvie accompanied him. One evening at dinner, Léonardon prophesied that El Dorado was at hand: a hundred crews, $40,000 in annual revenue. The following day, my mother took him aside and admonished him, "Don't talk to my husband about such fantastic projects; he could not sleep a wink all night."

The thought of such overwhelming expansion upset my father. Was that what he had been looking for in the cellars of the Ecole des

Mines, back in the years of his youth? Had he—an alchemist, though he did not realize it—had he suspected that the sand in which he planted his electrodes concealed nuggets of gold? The monotonous unfurling of the Texas landscape, seen from the train that was taking them toward California, added to his anxiety. The swampy forests and the blue-green Spanish moss that smothered the trees seemed to fill him with gloomy foreboding. He hid his troubled spirit by watching the passing scene.

El Paso, on the Mexican border, offered some diversion. The eye was captivated by the cactus, like gigantic church candles scattered over the purple desert. And at Tucson, Arizona, Indians from the reservations at San Xavier and Papago displayed their basketwork. The dry air, the brilliant light, and the rainbow reflections from the rocks quieted one's uneasiness. Finally, on a magnificent October day, Roger Henquet welcomed them at the Los Angeles terminal. Elegant, self-assured, radiating optimism, he swept aside shadows and obstacles with a wave of his hand.

On October 15, 1934, a letter was circulated to members of the international petroleum industry, informing them that the Schlumberger Well Surveying Corporation* had been formed. My father was chairman of the board, my uncle president, Léonardon executive vice-president.

Now, the skies once more serene, Conrad took his wife and daughter Sylvie on a voyage to the Hawaiian Islands, a place he had visited as a young man.

---

*Schlumberger Electrical Prospecting had granted S.W.S.C. the right to exploit its patents in return for the sum of $3,005,000, payable over fifteen years.

13

Counterclaims and Compromise

AFTER two decades, the idea which my father had planted in a crate of sand began to bear fruit. At the International Mining Congress in 1935, the Schlumberger methods and equipment attracted intense interest from a considerable number of specialists. "Germany is going well, Austria is developing, Italy is interested—the former Triple Alliance," Conrad wrote to me. On my wall map, the blue pinheads spread in all directions. The number of crews rose from 42 in 1933 to 74 in 1934, 127 in 1936, 237 in 1938. In America, once the Schlumberger Well Surveying Corporation was on its feet, Léonardon settled for good in Houston, at the economic heart of petroleum land.

In July 1935, he paid $9,000 for a rectangular tract of about 2.5 acres. "Léonardon has odd tastes," Conrad remarked. "He's the one who chose Houston. There isn't one of us who would go there to live."

Rarely had Conrad's foresight proved so mistaken. "I took a lot of blame for establishing myself on low ground," Léonardon tell us, "but when the low ground swallowed up the center of town and increased five hundred times in value, no one sent me any bouquets." Never one to lose his bearings, he had also suggested to *ces messieurs* that they invest in real estate. They let him talk. In 1938, he opened a three-story building on his land. Grouping together equipment service, truck assemblage, and a number of offices, the installation was worthy of his "troops": 106 engineers, 275 assistants, and 115 other employees. Not everyone applauded—not the white staff, because the area was black; not Paris, accusing Houston of tilting towards gigantism. Actually, a short year later, the whole building, containing seventy-two working areas, was fully occupied.

I note with some amusement that in a brochure commemorating

the second anniversary of S.W.S.C. and describing the Company's services with illustrations and diagrams, lateral coring stands out by its absence. This was not, however, for want of effort to commercialize the new invention. One day, when Marcel was leafing through a catalogue of the Manufacture d'Armes et de Cycles de Saint-Etienne, he came across the description of an underwater gun used in fishing, which gave him the idea for a tool that would bring underground samples to the surface. First tried out at Péchelbronn, the system — a gun firing three hollow bullets — was introduced in the U.S. by Louis Allaud; Jean Mathieu, then in charge of Gulf Coast operations, lost no time making it known to clients there. Since, in those days, no one thought of drilling without mechanically collecting cores, and because the new tool no longer took the cores at the bottom of the well but laterally in the wall, it seemed to promise the moon to many clients. Disenchantment was bound to follow; connections broke, bullets shattered, all sorts of mechanical difficulties prevented the apparatus from doing its best. A successful firing required particularly favorable conditions: The boring must not be too deep, the formations be easily penetrated, and so on. Allaud tells me about the expression on the face of a driller in Corpus Christi whom he had persuaded to try the operation:

> *I did my best to explain the maneuver to him: "You screw the gun onto the bottom of the sonde, send it down the well at the end of the cable, fire the hollow cylinders at different levels of the hole; each cylinder is attached to a wire, you recover the cylinders by pulling up the wire, and that's how you get your cores." He must have thought I was out of my mind. And at that, since I was using my best English—that is to say, talking by gestures and monosyllables—I probably did seem a bit nutty. Anyway, for once, everything worked smoothly and I still remember his surprise when I handed him two or three cores impregnated with oil. "The goddam Frog took a core!" he exclaimed, unable to believe his eyes.*

To my inexpert eye, the black box of my childhood had come to look like a strange electrocardiograph. Picking up waves and pulsations, it recorded the heartbeats of the earth. To anyone who knew how to read the peaks and troughs of the curves inscribed on the long paper strip, they represented an almost infallible geological record. *Wherever the Drill goes, Schlumberger goes:* This laconic slogan, coined by Jean de Ménil in the '40s, meant just what it said — no drilling without its electrical logs.

But our phenomenal rate of growth had stirred greed in many quarters, and, inevitably, attempts at imitation. Several court cases ensued. I shall mention only the two principal ones, for each of them in its own way shows how unassailable the Schlumberger position was, even though the second lawsuit was lost on appeal.

The first action was brought against Lane Wells, a California company that specialized in perforation. When a well passes through more than one reservoir, perforation makes it possible to put the reservoirs into production selectively. But the system does not function properly unless the levels of the reservoirs are determined with precision; this makes the well dependent on logging. In order to get around this difficulty, Lane Wells had acquired a Swedish patent, "the Geoanalyzer Company." Lane Wells, in his "lack of education in matters electrical," as Henri Doll said, "thought the Geoanalyzer was all the more sensational because the system was overly complicated." It was the validity of this patent that S.W.S.C. attacked.

The trial for breach of patent opened in March 1938 in Los Angeles, where I had gone with my husband. The history of the origins, workings, and patents presented by Schlumberger at the bar, the experts' depositions and the attorneys' exhaustive analysis of a documentation hard to refute, led the defendant to the conclusion that it would be better to come to an amicable agreement than to argue in vain. A triangular correspondence by telegram began between Los Angeles, Houston, and Paris. Léonardon was firmly against settling; he wanted Lane Well's hide. Henri was willing to enter into negotiation. Marcel, who had a dread of litigation, gave the green light. While the hearings in court ran their course, bargaining went on behind the scenes. In fact, since each party had its eye on the other's domain — Lane Wells on electrical logging, covered by Schlumberger patents, and Schlumberger on perforation, covered by Lane Wells patents — it became possible to find grounds for agreement. All this was long, drawn-out, and laborious. The meetings continued throughout the day, with a mediator presiding. In the evening, after a bite to eat, each camp worked out its tactics and strategy for the morrow. My husband came back to me late at night, completely worn out; his fatigue was such that when the time came to return to France he fell ill. As for the mediator, he died of a cerebral hemorrhage. He worked too hard.

The negotiations had ended in an exhange of licenses, giving each company the right to exploit the other's techniques. They were forbidden to establish competitive tariffs for similar services. The agreement, concluded for fifteen years but renewable by mutual consent at the end of five, covered only operations in the United States. S.W.S.C. was to pay Lane Wells a 12.5 percent commission on perforation, Lane Wells to pay Schlumberger 15 percent on logging. Besides, since Lane Wells had a smaller market for their services than Schlumberger, they were to pay the latter company the additional sum of $250,000.

The marriage now sealed and the signatures affixed, a banquet was in order, with cocktails, candles on the tables, orchestra, and photographs. There were six Schlumbergers, counting me, and thirty Lane Wells people. Hollywood palm trees, Hawaiian decor; the place was called Coconut Grove. I am afraid I didn't cut much of a figure but the others danced till dawn. Good humor was pervasive, friendliness demonstrative. The Lane Wells people were satisfied: They were sure they would take over the logging market while S.W.S.C. would lose out on perforation. What actually happened was the exact opposite.

About the same time, S.W.S.C. had to face a much more dangerous competitor. Halliburton, a powerful Texas enterprise specializing in well-cementing, attacked the logging market in the purest style of commercial rivalry — price wars, questionable deals, xenophobic arguments and what not. Halliburton had the support of Schlumberger's best client, the formidable Standard Oil Company, from which it held its technical weapon: exclusive rights to the Blau patent. This patent, developed by a researcher at Standard, was not invulnerable to legal action. Because the Blau impedance (a specious term, capable of misleading the judge, impedance and resistance being the same thing) used a single electrode instead of three — creating, by the way, inaccurate resistivity curves — it might possibly happen that the court would refuse to invalidate the patent. When a compromise was proposed, Halliburton refused; to offers to buy out their contract with Standard, they responded with offers to buy S.W.S.C.

After three years, Schlumberger, tired of the wasted effort, took Halliburton to court. I remember that the engineers' state of mind came close to exasperation. "We can't let ourselves be crushed," I heard it said around me, "by an outfit that doesn't know a thing about geophysics!" The trial before the Houston court lasted several weeks. As in Los Angeles, we had to educate the bench, repeat the history of the inventions, produce masses of documents, authenticate them, call witnesses, and so forth. The court ruled in S.W.S.C.'s favor on all points, invalidated the Blau patent, and prohibited Halliburton from using electrical logging. Halliburton appealed the decision.

It was now up to the Federal Court of Appeals in New Orleans to rule on the case. This time things did not drag on so long. Also, the scenario was not quite the same — no public audience, no hearing of witnesses, no arguments in form. The documents, minutes, and steno-graphic reports — enough to fill ten thick volumes — were deposited with the clerk of the court for the edification of the judges.

When their honors had absorbed the substance of this reading mat-ter, the attorneys for both sides came to sum up their theses. S.W.S.C.'s case was thrown out of court. The reasons adduced are of some interest. I extract these characteristic lines:

> *In our day the attitude of the public is hostile to monopolies, including those based on patents. The client tends to oppose them, preferring, it seems, to deal with more than one source of supply. It may be said that the temptation to break or circumvent monopolies is the greater because the footlights focus on a single monopolist who appears to have found his pot of gold.*

In one fell swoop, so to speak, the logging patent fell into the pub-lic domain, unless the case were carried to the Supreme Court — a long and risky procedure which assured neither the final decision nor even that the appeal would be received. Henri still remembers the day the New Orleans tribunal pronounced its decision. "At Schlumberger in Houston," he tells me, "everyone looked like a mourner at a funeral. I was about the only one not bothered. Rather than a disaster, I saw here a stimulant. Had I been Standard, I would have done as they did; I would have taken the trouble to finance a competitor. The American petroleum industry could not and should not have been at the mercy of a single supplier. At a given moment, we ourselves made great efforts so as not to depend on a single manufacturer of cables: He had us by the short hairs. As for logging, I was of the opinion that competition would keep us from resting on our laurels. It was up to us to do better all the time, and up to the clients to judge by their results. No, I wasn't bothered a bit. Besides, it was not the superiority of our technology that was at issue, it was our *de facto* monopoly. That is so true that once Standard had succeeded in breaking our patent by injecting Halliburton as a third party, they came right back to Schlumberger."

Henri's point of view won the day, and there was no third trial. Before long, S.W.S.C. had regained the ground lost to Halliburton — and more!

14

The War Years

UPON our return from Los Angeles in May 1938, it seemed to me that the face of France had changed. Those ten weeks of absence had made me more sensitive to the political situation. I sensed animosity in some people, and an undertone of hatred, a kind of resignation in others. To whom could I talk about my anxiety? I did not know. What would my father, concerned as he was with man and the human condition, have thought and done? His friend Jacques Michaut once told me that at Verdun, before Douaumont, Conrad had said to him, "You see, I'm ashamed of belonging to a family of textile manufacturers whose employees got up at dawn, walked for miles to reach the factory, and worked twelve or fourteen hours a day. Not one of those workers had a life fit to live."

Yes, Conrad's voice would have echoed mine. But Henri? . . . I picture him now as he was on the eve of the war and still is today. The first thing that impresses one about Henri-Georges Doll is his natural elegance, which shows in his bearing, his gestures, his long, knobby fingers—the hands of a craftsman in the noblest sense of the word. Surprise him in the field dressed in knickers, or at his work table in a tweed jacket, and what will strike you first is the ease with which he fits into his surroundings. One cannot imagine that a potentiometer would fail him, or that the lead would dare to break at the tip of his pencil. Around him and in him, but above all in his work, things are never hostile to Henri. He seems to be one of those men who could walk across the Gobi Desert without getting dust on his shoes; the passage of years would only accentuate his neatness and complicity with matter. It is impossible to know whether this surprises Henri: More likely, as his sparkling eyes suggest, it amuses him. Author of dozens of inventions and learned papers, he will explain his latest research with the same zest he had in the days of his earliest achievements. At

111

such moments, he is more youthful than a freshman. But the current event — History in action — has no hold on him.

Uncle Marcel was just as single-minded. Marcel liked mechanics — indeed, loved it exclusively. I always perceived him as incessantly adding refinements to whatever he was working on, asking nothing of anyone, wrapped in a silence that locked him in his own isolation. Weekdays he reigned over the drafting room. Sundays, tough and straight as an oak, axe in hand, he relaxed his muscles. His blue eyes, with their flashes of irony, put a brake to my enthusiasms. Behind them I felt a kind of shyness, a vulnerable sensitivity that I dared not intrude upon; to have referred to it would have been outside the rules of Marcel's game. His fine, classical features allowed one to glimpse the shy, withdrawn little boy he must have been. He never lost that child's charm and naturalness.

Henri, Marcel . . . that was how I saw them, that was how I loved them.

Schlumberger's international outreach now aroused concern about the status of the engineers at posts outside France. They were all of military age, could be drafted, and were therefore in danger of being trapped in enemy country if and when war broke out in Europe. Those in Germany were the first to be called home; then, as the situation worsened, the crews returned from Italy and Poland. There were still the neutral countries. In mid-April 1939, the missions in Romania and Hungary received instructions to sabotage their equipment if the need to escape arose. The remarkable thing is that they not only managed to survive throughout the war but also to supply Paris with currency. Poirault even found a way to obtain gold coins in Romania; he was able to pay his staff as well as to keep the equipment he happened to have — all of it pieced together on the spot — in working condition. As time went on, he fabricated his igniters out of tin cans and patched his cables with wire pulled off old rheostats. Scheibli, who is as hot-headed as Poirault is calm, struggled along the same way in Hungary.

To come back to 1939 . . . in early September most of the male personnel were ordered to report to mobilization centers. Some of them were drafted immediately, and later found themselves prisoners in the same camp (among these were Boris Schneersohn and René Seydoux); others, for some unknown reason, were granted postponements. Jean de Ménil, in Romania for a time, was mobilized on the spot; Eric Boissonnas, who also was there, was not mobilized, or at any rate not yet. On the other hand, each of our men overseas received his orders "to join up immediately and without delay," as if they were in the suburbs of

Paris. No one knew what to expect of the various French consulates with their jurisdictional prerogatives. In the U.S., engineers living in one state but working in another were claimed by two bureaucracies at once, with the result that the German legions were rolling through Chartres and the Beauce country while those concerned still awaited their marching orders. All of them stayed out in the end—some on the ground of special assignment, the others because the *Bretagne*, which they were supposed to board in Martinique, was sunk at sea. After the fall of France, some of these men were able to join the Gaullist forces in London and Africa.

The same men who just before the war were clients of Schlumberger now gave the company its orders. The attitude of Dr. Paul, who before the war had been a Schlumberger representative in Germany, was as cooperative as the new situation permitted.

Petroleum being the lifeblood of war, seizure of personnel and equipment by the Germans was a daily threat. This required continual improvization by the Schlumberger people still left in Paris. A strictly negative attitude toward the occupant would have done no good at all, so a policy akin to walking a tightrope was pursued until the end of the war. In the spring of 1940, Marcel had made arrangements to move *Pros*'s technical capital to an area where it would be safe. The thought of evacuating Paris was in the air. Most of the documents and drawings for equipment went to Clairac in southwestern France, but the especially important pieces, as well as the prototypes, were shipped to Houston. A truck equipped for logging—the only one then in France—was transferred from Péchelbronn to Saint-Gaudens, southwest of Toulouse. The truck included the spontaneous polarization dipmeter which Henri Doll had completed not long before, and the photoclinometer perfected under his direction by Bricaud, Jaeck, and Lebourg. This truck aided in the discovery of natural gas at Saint-Mart some time later.

A project for a mine detector to be mounted on tanks was also afoot. Raoul Dautry, who was then Minister for Armament and whose idea this was, called upon Schlumberger among others. The answer is said to have been, "Fine! But if you want to get off the ground you had better send for Doll; he will give it the necessary push." So Henri Doll, looking strange in his artillery lieutenant's uniform, came back and tackled the problem. He had in mind a kind of radar-before-the-fact, except that his device was intended to put the finger, so to speak, on targets very close to the vehicle carrying it. "I had come up with a few gimmicks," he told me, "but the advance of the German armies stopped all that and I had to go back to my regiment." He left behind a cumbersome model with Marcel, who was about to leave Paris with a heavy heart; he entrusted it to me to be put in a safe place. The mine detector was developed much later in Houston for the American army.

Before the war, research and manufacturing were concentrated in Paris. From there the crews, wherever they were, received their equipment and instructions. However, the invasion of France threatened to disrupt this system. Our laboratories and workshops were practically sequestered. To get them functioning again (even if that was possible at all), we were sure to be hampered by restrictive measures including rationing, requisitions, and export licenses. The mounting difficulty in making exchanges with the outside world would impose a closed and unavoidably less productive economy upon the crews. Some of them might manage to survive by prolonging the life of their materials, but it could not be the same for Schlumberger Well Surveying, Inc. There, shortcuts and tinkering would not be enough; the growing volume of logging in the U.S. could not continue in the absence of sources of supply outside France. S.W.S.C. had to have manufacturing shops adequate for its needs, and if it was to stay on course it needed an engineering service. Moreover, it was important to put in place a directorate whose functions would no longer be exclusively administrative (as was then the case), but technological as well.

"The force — the locomotive that pushes and pulls all the rest — is research," my husband pleaded. I knew the plea by heart: All his life, Henri would labor to maintain the primacy of research over finance. In the late summer of 1940 he and Marcel were locked in debate over these matters. Not that Marcel questioned the priority of research, but he was apprehensive about the consequences of an "American-style" development of S.W.S.C. He feared that once the idea of autonomy prevailed, the center of gravity would shift from Paris to Houston. Events were to show that he was right, but at the time his resistance was in fact a rear-guard action. My husband's arguments were irrefutable. If the future of S.W.S.C. was not to be jeopardized, the real problem was to decide who had the necessary knowledge and authority to take control. Excepting Marcel himself, this could only be Henri Doll.

So my husband, demobilized and provided with orders for a mission in North Africa, left around the middle of September. Jean Mathieu went with him. We were all to meet in Lisbon, where I with my three daughters and Paule Mathieu with her two children were to go via Spain. We sailed for New York aboard the *Excalibur*, an old rust-pot hastily refitted to carry human cargo. She was barely out of port when she began to pitch and roll abominably. For about 350 available accommodations there were a thousand souls on board. Most of the refugees were jammed into the holds. Only a handful of favored individuals occupied cabins; we were among them. Seen from our bridge, the crush on the deck below was not pretty to look at; no one could make a move without stepping over someone else. And from the deck there rose a murmur, a rumbling, in which twenty languages mingled. After twelve days of seasickness, Liberty, giving light to the world, saluted our ship as she came into sight. Never had a statue looked so beautiful to me.

We left New York on the fourth of November. Times Square was black with people: Roosevelt had just been reelected. A drawing room and sumptuous bedrooms were ready for us aboard the train to Houston. Henri was surprised that Léonardon, habitually thrifty as he was with the company's money, had done things on so grand a scale. As for my daughters, they had other things to be astonished about. Since the platform was level with the entrance to the car and they walked to the car on a red carpet, they wondered "if the train could really move, because you couldn't see the wheels." The journey, lasting two days and three nights, gave them time to ask a thousand questions, which I did my best to answer.

The train pulled into the Houston station on a hot, humid November day. "It's awful! I couldn't live here!" said my eldest daughter. "I pray to God to take me back to France," added my second, going her senior one better. "I don't want to see anything!" the third flatly declared. Léonardon, Charrin, and their wives, all smiling broadly, were waiting for us on the platform.

Now to find a house! I ran around the city in concentric circles. Colonial style, Tudor, Gothic, Spanish; turrets, rotundas, colonnades; frame houses, stucco, papier-mâché houses. I settled for one built of rosy brick, set back from the road in a residential district. A driveway—also in brick—led to it. The house had no particular charm, but it was spacious and had a typical Southern porch. Fine oak trees surrounded it, their branches hung with filaments of Spanish moss that swayed gently in the breeze. In a few days the dwelling was furnished—pale wood furniture, tableware from Woolworth's. Neighbors, friendly and helpful, introduced themselves—ah! you're French, poor France, the Eiffel tower, it's terrible, are you going to apply for American citizenship—offering to help, to show me around: The dry cleaner is on such-and-such a street, the doctor on such-and-such another, the milkman comes by at seven. Sam, the gardener, familiarized me with flora of this humid, semitropical climate. Grass, flowers, trees, shrubs — he referred to them as his friends: Moses, Elly-Agnes, St. Augustina. He talked to me about his father and mother, who had been born and died right there where he was raking my garden. He swung his arm in a vague gesture, as if to put his strong hand around this whole section of the city, not long ago a swampy woodland. Apparently, emancipation had not meant any great change for him, any more than it had for his grandparents. But Sam had no complaints; he answered my questions willingly.

I felt — and I always feel — a rush of sympathy for black people, a sentiment hard to describe because it is so confused. On Sundays I often went to their churches, which were built of rough-hewn planks. Most of them were indescribably poor; some were painted white, which by comparison made them look almost coquettish. The approaches were muddy in the winter and dusty in summer. Inside there were backless benches and a bare altar. God must be pleased with this plainness.

Like all American cities, especially those in the South, Houston was oversupplied with churches; at this time, in 1941, there were close to a thousand for a population of less than 400,000. None of them dated from the French or Spanish missions established in the region during the eighteenth century. All that was left from the time when Texas became the twenty-eighth state of the Union (1846) was a trading post (built to last) and two frame houses. Out of several ethnic quarters the only one that had survived, with its dances and music and traditional rites, was Frenchtown, peopled with Creoles from Louisiana who had come there in the 1920s. One could still see the streets named Adelia, Lelia, Deschaumes, Roland . . .

Not many adults, with the exception of blacks and Mexicans, were natives of the city. An absolutely Homeric chauvinism reigned there nevertheless: Being white and Houstonian made you *ipso facto* seven feet tall. Until recently the cotton capital, now the petrochemical capital, a great port open to the Gulf of Mexico although fifty miles from the coast, Houston thought of itself as the navel of the earth. I no longer remember who it was who once told me, "You see, the creation of the world and the solar system and all the rest, including the discovery of America, was foreseen and arranged by Providence for the one special purpose of fostering the economic rise of Houston." This was said only half in jest.

Across the front of a vast, ugly building on Leeland Avenue an inscription in tall blue lettering proclaimed the existence of the Schlumberger Well Surveying Corporation. Léonardon took considerable pride in his building: "one of the very first in Houston to be air-conditioned throughout," he pointed out.

Everything had happened so suddenly that S.W.S.C. was caught short. Léonardon's building alone measured up to the needs; very little had been done about the rest — raw materials, manufacturing, stocks, and above all, research. My husband threw himself body and soul into getting things moving. I frequently brought him a lunch box at noon — the kind used by American workmen — containing a frugal bite to eat which we shared at the corner of a desk. I listened as he talked about his problems. There would be the difficulty of getting along with some-one, or the trouble he had making them realize the importance of a project such as devising a new tool. "As for the technical questions," he said over and over again, "all right, it's up to me to make decisions; but I still have to beg and persuade and write reports — why do I have to waste my time? And they do things their own way without consult-ing me, and risk catastrophe." Paradoxically, Henri had against him the fact that he spoke for the absent Marcel, whose competence was beyond compare, who was his senior — in a word, who had too many trump cards up his sleeve. It was not that Henri wanted only to impose his own ideas; he fought on the ground of principle. I always saw him insisting stubbornly on the primacy of the researcher over the businessman.

In 1941, the Anthony Lucas Gold Medal was awarded to the Schlumberger brothers for their contribution to geophysics. There was a moving ceremony at the American Institute of Petroleum Engineers in New York, at which a scroll and a medal were given to me as the only representative of the family name in the United States. A number of speeches and a great deal of handshaking followed. To my surprise, no one mentioned the war: Europe, Africa, and the Far East were far away.

We returned to Houston; Henri plunged into his work, I resumed my various occupations. And before long, lightning struck the U.S. — the bombing of Pearl Harbor. In the following weeks a series of laws were passed governing priorities for the most "strategic" materials, including steel, copper, bronze, tin, rubber, jute, and cables. Gasoline rationing was also imposed, and drilling felt the restriction immedi-ately. The result was a sizable curtailment of oil prospecting and a cor-responding falloff in Schlumberger activity. "That was when we made a little mistake," Léonardon told me modestly. "The American govern-ment sounded the call for certain classes of specialists; we took the opportunity to send some people—even more than were called for." So that when the U.S. went full speed into oil production and drilling escalated like a shot, S.W.S.C. was short of engineers for the duration of the war.

15

The International Structure of Schlumberger

IN Romania, where Jean de Ménil toiled by day at *Pros*'s affairs and by night at those of France (with two of his colleagues, like him mobilized in Romania, he sabotaged the oil trains leaving for Germany by stuffing the grease boxes of the axles with iron fillings), the regime of "national rebirth" was visibly falling apart. In the days that followed the Franco-German armistice, the Soviets annexed Bessarabia and northern Bukovina. The Wehrmacht, under pretext of protecting the oil fields against nonexistent British incursions—but in fact to bar the Red Army's route—was infiltrating farther and farther into Transylvania.

The Bosphorus offered the only escape—a precarious one. Jean de Ménil and Eric Boissonas, my brothers-in-law, carrying out three automatic recorders and a packet of pounds sterling in a false-bottom suitcase, succeeded in getting to Turkey. Wasting no time there, Jean undertook to tour the *Pros* missions in the Far East, more than once barely eluding internment as he crossed frontiers. I remember his accordion-type passport, bulging with fifty or more leaves mottled with multicolored stamps, which he used to drop open like a Chinese lantern for our amusement. With this document in his pocket, he crossed the Pacific and landed in the United States two weeks before we did. It took Eric, who was in poor health, eight months to make his way to France.

Jean had married my sister Dominique in 1931. I was curious about this young man, who at twenty-seven was in charge of investment services at one of the largest French banks. After he became a member of the family he often heard us talk about *Pros*. To his ears what we had to say sounded vaguely unreal. One day (in 1934), he watched Conrad, face gleaming, exclaim at the news of an unusually ticklish operation, "We've got it! We're taking off! "The peaks and valleys of our enthusiasm showed him what he took to be a small family enterprise struggling to get along, not at all sure of its tomorrows, and kept afloat by a kind of visceral faith. It was all rather abstract — "a bit poetic," he said, "but extraordinarily attractive."

In January 1939, after much soul-searching, Jean joined *Pros*. Two reasons had kept him from yielding to the offers first of Conrad, then of Marcel. As I see it, his first reason was an idealistic view of the role he thought he might be able to assume as a banker. From a memorandum in which he analyzed the situation for Marcel's benefit, the motives behind his reluctance come out in these characteristic lines:

> *There are also—and I think this is what matters most to me—the 600 people I have under me. What a wonderful adventure it would be if I succeeded in establishing friendly relations with them, breaking through their distrust and resentment. The feeling of being one of the rare ones who approach this job with a new heart. Desire to have a part, there or elsewhere, in the quest for a harmony cruelly absent from the life of these great anthills. Now you will understand, I am sure, that if* Prospection *proposed a purely financial and fiscal post to me, I would be very reluctant to accept it.*

Concern for other people was an integral part of this man, whose emotions were so readily stirred. There was no pretense in his letters from Romania when he noted his regret at "not being able to share the lot of the poor guys the war has taken from their families," nor was there anything insincere in his avowal to his wife, who was coming to join him: "And here we are about to enjoy the priceless happiness of being together while others are fighting and you, because we are rich, can go traveling while others don't even dare think of it." Elsewhere he wrote, "I am horrified at the cowardice of the man who, at forty, sticks to his post but renounces the ideals of his youth."

In Jean's case, what exactly was that ideal? I do not think I'm wrong in suggesting that he had cherished the dream of a life dedicated to serving the weak and the rejected. A curious character! For in the end, however spontaneous may have been the movement that carried him towards others, he was nonetheless tainted with elitism. Jean believed in the demonstrative power of example, the duty of the "chief" to preach by example. Not that he was a giver of lessons or that he set himself up as a model. But strange to say, this man who wore his heart on his sleeve, this most brotherly of men, succumbed at times to what I am bound to call flashes of authority. His generosity, which his actions never belied, was yet at such times crossed with rudeness, even shattering harshness. He then was capable of making brutal demands upon his subordinates and peers. It seems to me that what unleashed these storms — as stunning as they were rare — was almost always a refusal to follow his lead on ground which he had long and painfully explored, and which therefore appeared to him to be the only right one. It may be that being of a visionary, distraught temperament astray in the world of finance, he feared nothing so much as failure. "I'm too attached to what I am doing, and once I've given myself to it I want it to succeed right away. This is so true that it's hard, if not impossible, for me to ask God to send me a failure if he thinks it's good for me."

The second reason for Jean's long delay in joining *Pros* was a matter of principle. He had won his position in the bank strictly on his own, but did he not owe *Pros*'s offer to special privilege? His code of ethics forbade him to lay claim to a high-ranking position because of his connection with the family. As he looked at things, being Conrad's son-in-law, far from guaranteeing his integration into the business, would make it more difficult. He could have made his own Gide's statement: "I have a horror of unfair advantage and do not wish to profit by anything that I have not earned by my own merit." Granted that Henri Doll and Eric Boissonnas also had married Schlumberger daughters, they had qualifications that Jean did not possess: He was neither an engineer nor a geophysicist. Their training was in line with *Pros*'s operation, so that in a sense their places were made for them; but it could not be taken for granted that Jean, an administrator by training, would fit in to the same degree. He had wide experience in business and had worked in a very large company at management level, but had no scientific experience at all. "All I know about electrical logging is how to spell it," he said.

The day came, however, when he had to give in to Marcel's insistence: the day Marcel, in his finest basso, grumbled, "Enough! We need you. You've got to come in," Jean understood that he could stay on the outside no longer. In the memo already quoted, Jean had asked to be allowed to spend some months in the workshops and in the field, since he refused to swell the "phalanx of industrialists who don't know how to hold a tool." At the same time, in order to show his colors for all to see, he declared that his coming to *Pros* must not create a precedent,

and that in no case would he consent to letting family interests inter-
fere with the progress of the business:

> *In my working experience, I have seen too many instances of businesses*
> *brought to ruin by a soviet of sons and sons-in-law, all extremely con-*
> *vinced of their right to equality, all fussing with everybody and giving*
> *orders left and right, the least experienced and the least active stifling*
> *the initiative and authority of the better ones. I think the Company has*
> *reached the point in its development where a formal structure is imper-*
> *ative. It is perfectly normal that members of the family be brought into*
> *the enterprise in preference to strangers, but only on condition that*
> *their place in the Company be in accordance with their value to it, and*
> *that their status as shareholders not give them the advantage of occu-*
> *pying posts for which they lack competence. The structure of the Com-*
> *pany must respond solely to its needs; no other consideration can confer*
> *the right to be a managing director.*

Jean de Ménil never budged from his position. In the course of
time it caused a bit of wrangling, and did much to bring about the
transformation of Schlumberger from a family-held business to a joint-
stock corporation.

Until shortly before the outbreak of war, the Société de Prospection Electrique had not given much attention to the problem of international payments. Primarily concerned with technology, the Company had allowed itself to become mired in a swamp of fiscal
regulations and exchange controls that varied from country to country.
Jean's first task was to clear the ground in Romania. There, the petroleum companies paid off their debts to Schlumberger partly in local
money, partly in hard currency. Since they kept their accounts of for-
eign-currency payments based on the unofficial (or black-market) rate
of exchange, whereas Schlumberger kept its accounts of receipts according to the official rate, an inextricable fiscal imbroglio resulted. Jean
spent a whole summer in discussions with the Romanian treasury offi-
cials and in negotiations with the oil companies to get them to admit:

a) that since Schlumberger was a foreign enterprise, its services
should be paid for in foreign currency in Paris, London, or New
York;
b) that a portion of the currencies would be reconverted into
Romanian money in order to take care of local expenses (salaries,
rentals, supplies, taxes, etc.);

c) that because this reconversion was made at the official rate of exchange, payments to Schlumberger in dollars, sterling, or francs could not be figured at the unofficial rate.

This agreement had hardly been reached when the war raised the whole question again.

Anyway, the lesson was not lost on that account; in fact, it turned out to be decisive for Schlumberger's development. No one could doubt that the aftermath of the war would make the problem of payments and transfer of funds increasingly thorny. Unless the Company's economic structure — which was completely unadapted to the existing situation — was remodeled, it was in danger of going down with all hands.

Jean had a foretaste of this as soon as he arrived in the Far East. The difficulties he ran into there, exceeding those encountered in Romania, showed him the extent of the danger and confirmed the urgent need to apply remedies. The Société de Prospection Electrique *(Pros)*, being a French enterprise based in France, was, under the Occupation, in enemy territory. This created a highly complicated problem (both commercial and financial) with regard to Schlumberger's dealings with the Allied powers. Jean was convinced that because of wartime conditions Schlumberger had to cut loose from *Pros*, and that the separation, at the moment a *de facto* reality, had to be made law. Ironically, Schlumberger's long-term survival was assured in the short term by the repudiation of *Pros*.

Jean tackled this complex transition with characteristic effort and perseverance. He organized his files by themes and interlocking series of propositions, each following as a consequence of the preceding one. To the building up of this mosaic, in which every piece had to fit perfectly or confusion would result, he brought a degree of concentration that on some days left him on the brink of exhaustion. Then his step was heavier, his gestures nervous, his speech more abrupt. But Jean knew how to take his time: The adage "make haste slowly" was made for him. He was to invest two years in convincing, coordinating, and putting in place a pattern of international structures which, in conjunction with S.W.S.C., played a decisive role in Schlumberger's spectacular growth.

I do not intend to review all the monumental effort to which Jean and his fellow workers committed themselves, except to say that, given the circumstances of the time, it was a superhuman one. Among the organisms created from the ground up, independent of each other and having no organic link with S.P.E., I shall mention only Surenco and Schlumberger Overseas. The latter, set up in the Far East, could not realize its full potential until after the war. The Japanese advance through Southeast Asia left only India as a field of operations, and there the volume of activity was sharply reduced.

Surenco, based in Caracas, where Jean and Dominique went to live in 1941, controlled all work in South America. The state of affairs in Venezuela was critical. Equipment was supplied piece by piece. Washington, insofar as it allotted quotas at all, gave priority to the American petroleum industry. Moreover, submarine warfare being what it was, tankers carrying Venezuelan petroleum to the U.S. sometimes went to the bottom; the same fate befell some of the small amounts of material that Houston managed to send southward. Local manufacturing was tried, but the available personnel was not equal to it. The oil companies, struggling with their own supply problems, began to curtail the practice of lending such items as housing, garages, and workshops. Like the treasury surplus (i.e., the total holdings in currency), engineers' salaries were, by common accord, held in a special account in British territory. Lack of funds became acute, creating the need to borrow, go into debt, and raise the price of services in order to recover the resulting expenditures: This caused major differences with clients. In the end, by freeing Schlumberger from its dependence on renting in the field, the very thing that at first looked like an insurmountable difficulty helped to release the company from any other forms of control by the petroleum powers.

The restructuring of the engineers' status also belongs to this period of crisis and adjustment. The enlargement of our geographical area and the coincidental dearth of specialists had modified the national makeup of the crews. The personnel, recruited in the U.S., Canada, Argentina, and Trinidad, was no longer exclusively French. The momentum begun in 1934 was irreversible. Problems of cohesion and cooperation had arisen among people of different languages, customs, and modes of living. A common denominator capable of damping latent or open friction was lacking. English, the universal language in the world of petroleum, had already been made obligatory at every level among the crews. But it had become clear to Jean that beyond national idiosyncrasies, the deeper source of potential discord was disparity of pay. Here were men coming from different countries and cultures, but often called upon to work side by side in a country foreign to all of them; it was not healthy that they should be paid according to norms that favored some over others. What was needed was a salary scale so drawn up that the base pay — with complementary allotments for living expenses, bonuses, insurance, and retirement figured in — would be equal for all who had equal qualifications, no matter what part of the world they worked in. Once this principle was enforced, the highest salaries then being paid served as a base for the upward revision of the wage scale, so that today Schlumberger engineers are among the highest paid anywhere. The consequence of this move was the establishment of harmonious relations in every division: personnel, equipment, and service. From Libya to Tierra del Fuego, from the offshore platforms to the jungles of the Amazon, the restructuring of the company's and the engineers' status, carried out by Jean de Ménil during World War II, made the motto *Wherever the Drill goes, Schlumberger goes* the sign of service well and faithfully done.

After a long illness, Jean de Ménil died in Houston on the first day of June 1973. "Only a handful of us know how much Schlumberger owes to him," wrote Jean Riboud. Nothing could be truer. There were very few who were able to see the man behind the outward appearances. One such was René Seydoux, who had teamed up with Jean in work and friendship. Nothing ever interfered with their collaboration, which lasted over a quarter of a century. Jean's penchant for the cutting retort was matched by René's cool patience. When he was present, contention seemed to peter out. Sad to say, René passed away that same June. The two men, so different in birth and background and so close to each other even in death, had in common the same love of life and mankind.

16

New Enterprises

NOT long after Pearl Harbor, Henri Doll began to think about resuming his study of the mine detector. Reluctant to burden S.W.S.C.'s budget, he founded Electro-Mechanical Research with André Istel's aid and offered its services to the American army. Francis Perrin, the future High French Commissioner for Atomic Energy, made his scientific knowledge available to E.M.R., and several Schlumberger engineers worked with the company.

John Bullington, one of the most touching personalities I have ever met, accepted the presidency of this enterprise. It was not easy, however, to get the green light for it. There was suspicion in high places about this rather foreign group of researchers who, without being asked, offered to invest money and labor in top-secret projects such as mine detectors and self-guided missiles. In fact, there were rumors abroad that S.W.S.C. had been taken over covertly by individuals shot in from France — in a word, by German emissaries. Letters patent (if I may use the expression) were granted only after indefatigable efforts by John Bullington, who had his own ways of getting to the War Department, and then only with the reservation that E.M.R. would be kept in ignorance of similar studies in progress elsewhere. Henri was always a glutton for work, but I do not think I have ever seen him throw himself into a project so strenuously; several of his colleagues almost broke down from overwork. High officers from all branches of service marched through, hurried and impatient, as if their gold braid were at stake.

Then one day, after months of continuous labor, a tank, escorted by motorcycle police with their sirens blaring, sped through the streets of Houston. Henri and a group of engineering officers clung to the tank in martial style. The vehicle, fitted with wheels with flexible spokes and loaded with detection circuits and tentacular apparatus, resembled

a monstruous insect out of science fiction. That was enough to drive
the spy phobia underground. . . . How do you like that? These Schlumbergers are into the secrets of the gods. . . . The amusing thing was that
at about the same time, Dick Tracy, a comic-strip detective with a big
nose and a square jaw, was shadowing a gang of unsavory characters
bent on kidnapping a foreign scientist who was putting the final
touches on — guess what — a self-guided bomb. So "realistic" was the
description of certain elements of the gyroscopic device in the comics
that Henri, who was working on the real thing, was very upset. Next
came the thought that Chester Gould, the creator of the strip, had a
pipeline into E.M.R. John Bullington, duly alerted, saw to it that the
bloodhounds were loosed on a different scent.

This friend upon whom we could always count, who always supported us, had made me love Houston and the vastness of Texas. John
Bullington's premature, tragic death from a heart attack left those who
knew him and loved him with a feeling of irreparable loss.

To get back to our tank with its motorcycle escort. . . . It had
reached a large field outside the city, where experiments proceeded on
mines buried there by the Army. Already at a high state of perfection,
stopping the vehicle dead when it came close to a mine, the system that
Henri had devised shortly underwent some remarkable improvements.
Circuits wound on frames of various shapes (round, rectangular, octag-
onal, etc.) "swept" the surface of the ground without touching it, setting off a loud signal at suspect points. To function well, however, these
feelers had to be exactly parallel to the surface. If any unevenness in
the ground tilted the vehicle — however slightly — nothing worked; the
least asymmetry called for constant adjustments that were both irksome
and dangerous. On top of that, the system was completely ineffective
over magnetic ground. Such terrain was a peculiarity of the Japanese
coast, and also very common on the roads of France, which were paved
with granite over a base of coal cinders. These defects, which made the
use of detectors unsafe, had been eliminated by Henri's invention.

Not long after the Liberation, the War Department offered the
French Army the right to copy Henri's apparatus to help in locating
mines; Henri arrived in Paris to demonstrate its use. A plot of land sat-
urated with magnetite was prepared at the Fort of Ivry, ouside Paris,
and all sorts of detectors — American, English, French, German — were
tried out there. The E.M.R. device outclassed its numerous competitors,
easily overcoming all obstacles. Moreover, jeeps had already been
equipped with it at the time of the landings in Italy.

A few days before the Japanese surrender, Washington wanted to know whether the detector would function at a depth of a hundred yards, and if so, whether it could be made available immediately or sooner. Henri and Lebourg went right to work experimenting in one of Schlumberger's test wells, not taking time to eat or sleep; in twenty-four hours a fighter plane took off from Dallas with two directors aboard. Mission classified. . . . The key to the mystery was as follows: When the Philippines fell, General MacArthur had sacks and more sacks full of gold coins dumped into Manila Bay; they could not be fished out after the victory because the said sacks had desintegrated due to the action of seawater. We learned that the first thought had been to send Henri. However, since he was a foreigner, bureaucratic red tape would have required the intervention of eleven different departments. As things turned out the detectors did wonders and, once the treasure was recovered, Henri was entitled to a Certificate of Appreciation "for several basically new principles for the location and detection of mines and similar objects." I suppose he might have preferred a nice gold piece in place of a scroll!

Electro-Mechanical Research's accomplishments constituted a kind of moral victory, besides earning the patents to various systems which the American army, in consideration for the company's work, relinquished. One of these systems, called "phase selective feedback," became a determining factor in perfecting the induction sonde. The development of this tool was well advanced at the time. Several companies — each claiming the original idea for it — eventually went to litigation, but Henri was in fact its inventor, as was proved by a review of the confidential reports he had written for the research engineers at the War Department.

When the rapid reconversion of the American war economy to a peace economy brought a halt to E.M.R.'s activity, Henri, who saw in it the seed of a research center, vetoed its out-and-out liquidation. The company went on with studies and spectrometry, telemetry at the bottom of boreholes, and so on. Frequency-modulation telemetry later found practical applications in satellites and space probes. Transferred from Houston to Ridgefield, Connecticut, and then to Sarasota, Florida, E.M.R. equipped with digital telemetry the Apollo capsules that landed on the moon. Many hopes were vested in the small company, which was (relatively) master of its own program and freed from the demands of profitability. By absorbing E.M.R. Schlumberger took the first step toward extending its field of operation beyond the boundaries of petroleum prospecting.

With the end of the war, the question of a research center became more pressing than ever. Ridgefield, Connecticut, was chosen for the center's location. A large, wooded section of land on the outskirts of the little town was acquired and promptly cleared; in 1947, the offices and laboratories, designed by Philip Johnson, a leading architect, were under construction. "Research and Development," as the center came to be called, was the first of eighteen such establishments now scattered throughout the world. It was operational in 1948, the year Eric Boissonnas became its administrative head, while my husband retained management of its technical side. The two brothers-in-law complemented each other in a perfect entente. At the time of this writing, more than two hundred engineers, mathematicians, and physicists work there full time.

I had hoped that when the prophetic number of seven years had elapsed we would be on our way back to France, but Henri objected that his work would not allow it. Schlumberger, Surenco, and Schlumberger Overseas, he explained, were in the midst of administrative, industrial, and—most important—technical reorganization. Now that the Research Center had finally come into existence, he could not decently sidestep the duty of breathing life into it. Who else but he could put it on track? Where else but in the U.S. could the men, the materials, and the most advanced electronics be found? Surely not in France, which was then at the bottom of a postwar slump. Voluble and eloquent as he was, American technology and know-how moved Henri to lyrical effusions. All he asked was two years, three at the most, to get the machine on the road and up to speed. And besides, didn't New England, with its lakes and woodlands, remind me of Normandy without the rain? Though neither charmed nor convinced, I had to give in.

It was not that America left me indifferent. Through the Americans I knew, I was deeply aware of the still-recent epic story of a vast, wild country. There, time—and I mean History—took on a special quality: The visible and the tangible took shape before your eyes. It seemed to me that in this part of the world, what elsewhere took centuries was accomplished in the space of a generation. In a sense, as is said of a well-done work of art, in America everything is precision and energy. Success—the much-talked-of "opportunity"—is nothing but work and more work. Here, the rewards of personal merit, largely mythical though they be, still are less fictitious than in old Europe.

Thus, to take one example, Elliot Johnson, who joined Schlumberger in the '30s—the darkest days of the Depression—had his new law degree as his only qualification. What impresses me in his case is not his success (becoming the chief lawyer of the Company), but the fact that, starting from zero, he represents an upward social movement which, though not open to everyone, is still accessible to many. He was descended from Norwegian immigrants who came over in the middle of the nineteenth century — a long voyage in a sailing ship, where only the strongest survived. And the trial was only beginning. America was not merely a dock where passengers landed; it started in a fabled West, where green pasture awaited the first comers. Months of travel on foot, without nourishment worthy of the name, with no aspirin to calm fevers; long trails bordered with graves; fear of Indians and snakes; the setting sun their only road map. Then came a clearing, some woods, a stream, and expanses where the imagination saw a sea of corn growing. Their first dwelling was a hole dug in the earth and covered over with tree trunks. Everything was piled into it — adults, children, tools, seeds for sowing. Elliot talked to me about it as he had heard his grandparents talk — a story like thousands of others . . . like Rieke's, who was at the top level of management in Houston, and whose grandparents followed the tracks of the Transcontinental Railroad until exhaustion stopped them. Peasants from the remotest parts of Europe were the builders of this country, and their descendants sent their sons to the great universities. Oh, I knew that the Indians had suffered terribly, the blacks had been enslaved, and there was the blood and sweat of human exploitation in the country's makeup. But the Spanish seizure of South America was no more humane. The Old World's past is hardly more edifying. No indeed, I was not devoid of feeling for this melting-pot of immigrants, where eighty ethnic strains merged to form one nation.

17

From the Magic Box
to Aladdin's Lamp

AFTER seven years of Texas came the years in Connecticut, which were supposed to be "three at the most" but rose to four, then six, then seven in their turn. Was I destined to see my life measured in seven-year cycles? One can make numbers say anything, and the number seven does so — more than seems reasonable: There is even a certain encyclopedia that lists, between the seven stars of the Pleiades and seven-league boots, three-times seven wonderful examples of sevenness for your amused meditation. Anyway, I have indulged in a little mental arithmetic, and, upon my word, things drop in as neatly as you please: 1913, full-scale surface prospecting; 1920, the Schlumberger brothers begin to work together; 1927, first electrical logging; 1934, birth of the Schlumberger Well Surveying Corporation; 1941, Houston; 1948, Ridgefield, 1955?* ... I could run the enumeration-by-sevens quite another way, making it coincide with my father's experiments in the baby's bathtub, the Armistice, Conrad's first trip to the U.S., mine to the U.S.S.R., World War II, the unblocking of monies withheld during the hostilities, Marcel's death ... games for an idle mind, indeed.

During the autumn of 1951, it was my joy to have Marcel at Ridgefield. We had long conversations. He seemed more pensive, more careworn than I had remembered him, and his questions sounded more direct — I might say more intimate — than usual. I can still hear him saying in a low voice, "You ought to go back. You're too alone here. I'll speak to Henri." Once he asked me about Jean Riboud. "You see him now and then. What do you think of this lad?"

---

*It is interesting that in 1962 (seven years again) Schlumberger was listed for the first time on the New York Stock Exchange.

I answered that I held him in high esteem. "As you know, I regard him as a friend. I think he has a heart — a feeling for humanity, I guess I want to say. That's rare enough in someone committed to high finance. If you're thinking of taking him on, I'll be surprised if he disappoints you."

"Oh," Marcel said, "I find him *sympathique*. We'll see." Then, after a pause, "I wouldn't know how to use him. Finance is not our business, and I don't believe in it."

"Yes, but the way the business is growing creates problems. . . ."

"Maybe, but after all we're engineers, you know. It's technology, not finance, that concerns us."

In the half-light, as we waited for dinner, the talk went on in lowered tones. "I know," I said. "Except that these days, the future is being built on a wholly different scale. We can't help it. . . . Already, there's very little left in the industry that resembles what you and Conrad knew or even perhaps wanted."

Marcel did not reply. Was he thinking, as I was, about the thanklessness of the things one creates from nothing? They expand, grow tall, you think of yourself as the master builder, then one day the thing leaves you as ripe fruit drops from the tree that bore it.

One year later, Jean Riboud joined Schlumberger. Installed in one of the offices at the Rue Saint-Dominique, he learned the working of the enterprise with Marcel as tutor. I know that Marcel felt that his end was near and that his death might precipitate an outbreak of discord that would bring down the work to which he and his brother had given their best. To take Jean Riboud under his wing, form him in his school, and bequeath to him the stamp of his authority, was, to his thinking, the way to avoid the worst.

In April 1953, on a forty-five acre plot, S.W.S.C. inaugurated its new quarters in Houston. Eight thousand guests from the petroleum world, glasses in hand, roamed through offices, laboratories, shops, testing stations — 90,000 square yards of useful surface. Marcel was not present. Four months later, in August, he had joined his father, whose vision had preceded and often surpassed that of his sons.

Jean Riboud was ready to take over the leadership of the business.

Since 1953, the spirit of Conrad and Marcel has been felt and kept by each individual, even by those to whom Conrad and Marcel were only names. It is indeed extraordinary that the company has been able to maintain to this day this human touch and personal approach, despite its spectacular growth.

Marcel's fears came true earlier than he thought. With him gone, each person was aware of the responsibility and felt that he was the only one who could take charge of the Company. All had good reasons to change what they had inherited. The reality of *Pros* became only an image.

How and when did this desire to break with the past come to me? I really do not know. One day I tore up my roots and started a new life.

It is amusing to jest about the *Vert Paradis des Amours Enfantines.* Ah yes, the green paradise of the loves of childhood. . . . As is only right, the shine of my own paradise did not withstand the passing of the years. Impassioned witness of a long and patient effort—that was what I loved, not its spectacular success. As I write these lines, the two rooms at the Rue Fabert have spread out across oceans and continents. Implanted in seventy countries and operating eighteen research centers, the little enterprise of the '20s has become a multinational corporation, listed on the stock exchanges of New York, Paris, London, and Amsterdam. Its monetary value runs into the billions. Schlumberger is no longer just a trade name: Of the 85,000 people the company employs, not one bears the family name. At least in these pages, I have sought to tell in my own words how Conrad and Marcel's modest black box evolved into Aladdin's lamp.

Appendix

Schlumberger Limited Today

With 85,000 employees working in 78 countries and gross revenues in 1981 of nearly $6 billion, Schlumberger Limited is a large, international organization. It operates in two allied fields: One segment offers services to the oil exploration and production industry; the other segement is involved in the manufacture of electric and electronic measuring equipment such as meters, control instrumentation computer components, and computing programming. With both groups engaged in various aspects of measurement and in the manufacture and use of instrumentation, the two segments have much in common.

Electrical measurement, as applied to the new science of geophysics, gave Schlumberger its start in 1912. Conrad Schlumberger, then professor of physics at the Ecole des Mines in Paris, France, conceived "electrical prospecting" as a major advance over gravimetric and magnetic methods then widely used in exploration for minerals. Early experiments in 1912 and 1913 showed that the method was indeed feasible. Resuming work in 1919 after the end of World War I, Conrad was joined by his brother Marcel in starting a small company offering surface geophysical services. The firm, *Société de Prospection Électrique*, successfully located ore bodies, mapped buried geophysical structures, and located dam sites. By 1926, it was already an international firm with operations on all continents.

In 1927, the Schlumberger brothers applied their method by drilling boreholes for oil at Péchelbronn, France. The results were outstanding and led them to a new industry, the logging of oil and gas wells worldwide. This activity became so large that it was to replace their business of surface exploration entirely. Today, under the general heading of Wireline Services, the modern generation of logging methods is as vital to the exploration and production of oil and gas as the X-ray is to the practice of medicine.

Geologists believe that oil and gas (hydrocarbons) are formed by heat and pressure from the remains of ancient plant and animal life. Hydrocarbons thus created migrate through layers of porous and permeable rock until trapped by a layer of impermeable rock. Oil companies search for these traps using surface geology and geophysical tools such as seismic surveys. When the presence of a trap—and possible oil deposit—is suspected, a well must be drilled to see if hydrocarbons are present. But even then, a geologist, at the top of a hole a few inches wide and thousands of feet deep, possesses little information on the formations through which the drill has passed. To get this information, most companies call Schlumberger or another wireline logging company.

Oil-well drilling is a continuous operation. When logs are needed, a mobile logging unit is dispatched to the well. Drilling is stopped, the drill pipe removed, and the logging tools lowered into the well on an armored electrical cable called a wireline. As the instruments are retrieved, they measure in passing the depth and physical properties of the various formations pierced by the drill. Signals are transmitted through the cable to the surface, where they are recorded on magnetic tape and on a graph called a log. Interpretation of these logs yields information on the location and producibility of hydrocarbons.

There must be no loss of time, so Schlumberger maintains its mobile laboratories as close as possible to drilling operations wherever they may be. When drilling takes place at sea or in isolated jungles, a logging unit and crew may be maintained at the well site. This need to follow drilling operations closely explains why the company operates is so many places and different countries.

To keep logging crews supplied with equipment, Schlumberger maintains large manufacturing centers in Houston, Texas, and in Paris. To improve the quality of measurement and develop new, better methods for serving the oil industry, the company also maintains large research and engineering facilities at Houston, Paris, and Ridgefield, Connecticut.

Typical borehole measurements are based on many parameters such as electromagnetic, acoustic, and nuclear effects:

Electromagnetic   This family of tools measures formation resistance to the passage of an electrical current. As hydrocarbon-saturated rock is more resistive, these logs serve as a basic measurement for locating oil and gas.

Acoustic   The sonic tool transmits a sound wave and measures its travel time through the formation. This measurement is used for evaluating the porosity of a formation as a means of calculating how much fluid it might contain. Sonic logs are also useful in understanding certain mechanical properties of the rock and for information vital to interpreting seismic data in determining future well sites.

Nuclear   This family of tools employs natural or induced radiation to investigate the atomic and nuclear structures of matter within the formation. They help determine the rock type, porosity, and fluid composition of the formation. Several types of measurements are made. One tool detects natural formation radiation, which is helpful in locating shales. Another irradiates the formation with gamma rays as a way of measuring rock density. A third tool employs neutron bombardment to measure hydrogen content.

Many other wireline tools are used in the open borehole to obtain different types of information. The dipmeter, for example, provides information on the dip and strike of formations pierced by the drill as an aid to future well site selection.

Once the presence of hydrocarbons has been established, standard practice is to case the borehole with steel pipe which is cemented into place. At this point, another family of wireline tools capable of operating within the cased hole are required. These are needed to complete the well as a producer.

Major completion services include: measurements to determine the effectiveness of the cement bond that holds casing in place and isolates the reservoir from nonproductive formations; perforations of the casing at reservoir depth to allow oil and gas to flow into the wellbore; and nuclear logs for evaluating the formations behind the casing.

Once logging data are acquired, they must then be interpreted and analyzed. In the early days, this was done manually by logging engineers who performed computations on the spot—a time-consuming process. Today, data are fed directly into a computer on board the logging unit and interpretations are performed with great speed and accuracy for "quick-look" interpretation. Logging data from the well site may also be radioed to field computing centers for in-depth analysis to refine the accuracy of the formation evaluation. These analyses are used to engineer the testing and completion programs that follow. The third level of data processing and interpretation offers even more detailed and comprehensive study of logging data. Services such as Reservoir Description and Production Management logs are produced. Reservoir Description is an extended computer analysis which combines and compares logging data from all the wells in a given field together with other data (such as core analysis and production test results) to establish a comprehensive, three-dimensional picture of the oilfield as a whole.

As Schlumberger developed worldwide contacts in the oil industry, it was natural for it to diversify into other oil operations. Through Forex Neptune, the company provides contract drilling services in Europe, Africa, the Middle East, the Far East, and South America, operating 56 land rigs and 17 offshore rigs. Another company, Dowell Schlumberger, offers the important services of cementing and acidizing in the Eastern Hemisphere and Latin America. Johnston-Macco and Flopetrol provide such services as drill stem testing, production testing, reservoir monitoring, and a variety of production and well-maintenance services. The Analysts offers computer analysis of drilling operations and allied drilling support services.

With Schlumberger's experience in designing and building electronic tools that could withstand harsh environments and severe operating conditions, it was natural for the company to diversify into the electronic industry. The first of these ventures came through Electro Mechanical Research, a company created during World War II to do classified research for the United States Armed Services. Building on this base, Schlumberger expanded its electronic operations into several important fields. These now constitute a major segment of the company's business under the heading of Measurement, Control & Components.

Measurement, Control & Components comprise companies involved in electricity management and measurement, data systems, industrial controls, semiconductors, other electronic components, automatic test equipment, and computer-aided design and manufacturing services. The companies in this segment provide products and services to producers and distributors of electricity, water, and gas, as well as to aerospace, electronic, chemical, and food industries. They are divided into four groups: (1) Fairchild, (2) Measurement and Control – Europe, (3) Sangamo-Weston, and (4) Computer Aided Systems. The four groups are made up of 17 autonomous divisions spread throughout North and South America, Europe, and Asia. Each group has its own management, research and development, and production operations.

Fairchild's Semiconductor – Analog & Components division makes discrete components such as transistors and diodes, telecommunication products, and optoelectronic devices such as fiber-optic couplers. The Semiconductor – LSI Products division makes Large Scale and Very Large Scale Integrated circuits such as microprocessors, memories, logic circuits, gate arrays, and charge coupled devices using MOS, advanced bipolar, and CMOS technologies. The Automatic Test Equipment division makes computer-based systems for testing semiconductors, printed circuitboards, and subassemblies.

Measurement and Control—Europe includes the following companies and divisions: Enertec makes meters and load management equipment for electricity distribution, relays and transformers for electricity transmission, instruments and systems, data acquisition equipment, and magnetic tape recorders. Flonic makes gas and water meters and electronic payment systems. Sereg makes industrial control equipment and valves for a wide variety of industrial uses. Service offers services related to water and gas distribution. International makes electricity, water, and gas meters and related systems in several countries of Europe and Latin America. United Kingdom makes electricity meters, aircraft and industrial instruments, electronic instruments, training systems, and automatic test equipment.

Sangamo-Weston's divisions include the following: Data systems makes data acquisition, telemetry, supervisory control systems, and magnetic tape data-recording systems. Rixon makes modems and associated products for data communications. Electricity Management makes watthour meters and equipment for electric power distribution systems. Capacity makes capacitors for both electronic and electric power applications. Fairchild-Weston Systems makes optical and electro-optical data acquisition equipment and signal processing systems for aerospace and defense applications, and also controls for nuclear power systems. Instruments makes scientific and aerospace instruments, vehicle performance recorders, and photoelectric devices.

The Computer Aided Systems group includes two companies: Manufacturing Data Systems and Applicon. The former provides computer-assisted software services for numerically controlled machine tools and other specialized computer services for manufacturing industries. Applicon offers both products and services used in computer-aided design of products ranging from integrated circuit chips to complete automobile parts.

From an idea tested in a baby's bathtub in the basement of a college has sprung a major corporate enterprise; from the vision of Conrad and Marcel Schlumberger, who saw the possibilities inherent in applying electrical measurements to geophysical operations, has evolved a prominent organization that continues to grow and prosper. Schlumberger Limited, today, is a fitting memorial to the genius and wisdom of the men who gave it their name.

The success of Schlumberger has been largely due to its belief—

originated by the founders—that research and engineering are the life-blood of any enterprise. Readers who are interested in learning more about this side of the story should read the excellent book *Schlumberger—The History of a Technique* by Louis Allaud and Maurice Martin. The authors start with the early experiments of Conrad Schlumberger in 1912 and then trace the development of electrical surface prospecting from 1919 to 1932. They describe the development of logging from the first simple electric log in Péchelbronn in 1927 to the highly sophisticated techniques developed later; they end their story in the 1960s with the introduction of the computer that was to cause another revolution in oil exploration.

WILLIAM J. GILLINGHAM

Houston, Texas

Index

Academy of Science, Leningrad, 66
Aktiubinsk, 78
Algeria, prospecting in, 33, 95
Allaud, Louis, 98, 106
Allégret, André, 30
Alsace, 4, 5, 17–20
  salt dome, discovery of, 17–20
Amazon, prospecting in, 125
American Institute of Mining
    Metallurgical and Petroleum
    Engineers, 49
America, prospecting in. *See*
    United States
Anahuac, 98
Anglo-Iranian Oil, 86
Anthony Lucas Gold Medal, 117
Arizona, 103
Arts et Métiers, 34
Ashkhabad, 78
Asia, prospecting in, 85–91, 123, 124
Assam, 90
Assam Oil, 87
Astrakhan, 76
Autric, 13

Baboin, Guy, 13, 90
Baku, 49, 62, 65, 74, 78
Bandau, 88
Baron, 13, 15, 27
Batavia, 88
Bayle, Pierre, 35, 79, 81–83, 84
Bayou Serpent, 22
Beaufort, 85
Beaumont, 98

Bégin, 88
Bengal, 88
Benisaf, 33
Bessarabia, 119
Blau impedance, 108
Boissonas, Eric, 13, 112, 119, 121, 130
Bokhara, 78
Bolshevik Civil War, 78
Bombay, 89
Bor, 3
Boreholes, measurements in, 25–31, 47–48
Borneo, 88
Bourumeau, 37
Breusse, 13
Briceno, 83–84
Bordat, Louis, 56, 61, 84, 86–87, 90, 91
Brahmaputra, 88
Bricaud, 113
Bukovina, 119
Bullington, John, 127, 128
Buenos Aires, 89
Bureau of Geophysical Research, 80
Burma Oil Company, 86
Burma, prospecting in, 86, 87, 88, 91

Cabimas, 82
Calcutta, 88
California, prospecting in, 23, 48, 81, 93, 95, 102, 103, 107
Calvados, 3
Canada, prospecting in, 30–31, 93

Cape Horn, 89
Caracas, 81, 124
Casing perforation, 84
Castel, Jacques, 78
Catt, Mrs. Carrie Chapman, 52
Centrale, 34
Cégésec, 97
Charrin, Paul, 21, 37, 70, 76, 78,
　　79, 115
Clairac, 113
Clairin, René, 19, 37
Le Chemin de la Vie, 73
Colmar, 18
Commissariat of the People for
　　Heavy Industry, 74, 80
Compagnie Générale de
　　Géophysique, 94–95
Coring. *See* Electrical coring and
　　Logging
Corpus Christi, 106
Correlations, 82, 83, 86
Coste, 13

Dakhnov, 66, 72
Dautry, Raoul, 113
De Geffrier, Raoul, 37, 89–90
Delord, 13
De Ménil, Dominique, 1, 3, 35,
　　119–121, 124
De Ménil, Jean, 13, 106, 112,
　　119–125
De Ojeda, Alonso, 82
Deschâtre, Gilbert, 15, 21, 45–49,
　　81–83, 96, 99
De Witt-Schlumberger,
　　Marguerite, 5, 51
Digboi, 87, 88
Dipmetering, 85, 113
Doh, Charles, 85
Doll, Henri, 17, 25–27, 29, 35, 41,
　　43, 48–50, 52, 96, 107, 109,
　　111–112, 113–117, 121, 127–
　　130, 133
Donets Basin, 74

Dossar, 78
Douaumont, 111
Dutch East Indies, prospecting in,
　　81, 88

East Indies, prospecting in, 81, 88
Ecole des Mines, 1, 2, 12, 17, 18,
　　34, 38, 53, 102–103
Ecole des Sciences Politiques, 43
Ecole Polytechnique, 17, 34
Ecole Supérieure d'Electricité, 34
Electrical coring, 8, 12, 27, 29, 42,
　　47–48, 74; *see also* Logging
Electrical logging. *See* Logging
Electrical prospecting, 2–3, 12; *see
　　also* Electrical coring and
　　Logging
Electro-Mechanical Research,
　　127–129
El Paso, 103
Emba, 74
Equipotentials, 9

Far East, prospecting in, 85–91,
　　123, 124
Fierville-la-Compagne, 3
Five-Year Plan, 73
Ferghana, 78
Franco-Prussian War, 4
Frequency-modulation telemetry,
　　129

Gabon, 95
Gallois, 13, 15, 27, 48, 83, 99
Galvanometer, 101
Ganhati, 88
Geoanalyzer Company, 107
Geological Petroleum Research,
　　68
Geophysical Abstracts, 72
Geophysics, 11, 13
Germany, 112, 113
Gillingham, 99
Glavneft, 79, 80

Glutchko, 61, 67
Golubiatnikov, Professor, 53, 68
Gomez, 84
Gould, Chester, 128
Gravimetry, 12
Grigoriev, Professor, 61–62, 67
Grozny, 49, 56–60, 62, 65, 74, 76, 78
Groznyeft, 79
Gubkin, 72
Gubkin Institute, Moscow, 70
Guebwiller, 4, 5
Guichardot, Georges, 78
Guizot, François, 5, 51, 52
Gulf Coast, 95, 98, 106
Gulf Oil Company, 21, 48, 83, 93
Guriev, 78
Guyod, 79
Gypsy Company, 48

Halliburton, 108–109
Hand recorder, 84
Henquet, Roger, 13, 99
Hettenschlag, 18
Houston, 27, 96, 102, 105, 108, 109, 113, 114, 115–116, 124, 125, 127, 128, 133, 134
Hungary, 112

Ice, perpetual, 74
Imphal, 90, 91
India, 88, 89, 91
Indochina, 89
Indonesia, 89
Induction sonde, 129
International Mining Congress, 105
International Petroleum Company, 30
International Woman Suffrage Alliance, 51
Irkutsk, 79
Irrawaddy River, 86, 91
Istel, André, 127
Italy, 112

Jabiol, 88
Jaeck, 113
Jakarta, 88
Jaluca, 49
Java, 89
Johnson, Elliot, 130–131
Johnson, Philip, 130
Jost, 13, 70

Kakemyo, 91
Katanga, 34
Kazakstan, 78
Kelly, Sherwin F., 12, 30, 42
Kokand, 78
Kokubu, 91
Krasnovodsk, 78
Kuwait Oil, 86

Lahaye, Georges, 37
Lake Maracaibo, 83
Lalande, 35
Laguilharre, 90
Landes, 33
Lane Wells, 107–108
Lannuzel, 56, 58–59
La Rosa, 82, 83
Lateral coring, 106
Laubereaux, 13
Lebourg, 113, 129
Leleu, 89
Leninabad, 78
Léonardon, E.G., 13, 22, 27, 28–31, 41–43, 93–94, 96, 98–99, 100, 102, 103, 105, 107, 115, 117
L'Humanité, 71–72
Libya, 125
Logging electrical, 81–85, 95, 97, 100, 106, 107, 108, 109, 121
Louisiana, 99
Louis Phillipe, 51, 52

MacArthur, Gen. Douglas, 129
Magwe, 86, 90, 91

Makhumud, 77
Manipur territory, 90
Manufacture d'Armes et de
    Cycles de Saint-Etienne, 106
Maracaibo, 49, 81–82
Martin, Maurice, 73–74, 76, 78
Matamoros, 100
Mathieu, Jean, 79, 83, 95–96, 99,
    102, 106, 114
Mathieu, Paule, 114
Maylasia, 88
Dr. Mekel, 15
Mekteb, 77
Melikian, Vahe, 53, 57–58, 61, 63,
    67–70, 79, 80
Mennecier, 13, 84
Meny, Jules, 12
Mexico, 100
Michaut, Jacques, 111
Migaux, Léon, 80
Miller, Jules, 37
Mine detector, 113–114, 127–129
Dr. Moineau, 11, 15
Morocco, 95
Moscow trials, 71
Mulhouse, 18
Murray, George, 84
Musée d'Histoire Naturelle, 80

New York, 41–44
Nisse, Robert, 23
N.G.R.I., 74, 79
Normandy, 3, 33
Novoalekseyevka, 78
Nyangla, 87

Offshore measurements, 84
Oklahoma, 45–46, 48–49, 81, 93,
    98
Old Ocean, 98
Opuchinin, Mikhail, 67

Palembang, 88
Palustron, 89

Parbatipur, 88
Dr. Paul, 113
Paulin, Elie, 83
Péchelbronn, 25–26, 34, 49, 81,
    106, 113
Perforation, 107
Perrin, Francis, 127
Peshawar, 88
Petroleum prospecting. *See*
    Schlumberger
Phase selective feedback, 129
Philippines, 89
Photoclinometer, 113
Piotte, Roger, 35, 83, 84
Pladiu, 88
Poirault, 13, 35, 56, 112
Poland, 112
Polarization, 9
Poldini, Edouard, 12, 13, 28
Potentiometer, 34, 36, 49, 111
Pros, 22, 27, 33, 37–39, 41, 47–48,
    49, 54, 70, 71, 72, 73, 74, 75,
    80, 82, 83, 88, 93, 94, 102,
    113, 119, 120, 121, 123, 135
Proselec, 35, 93, 96–97
Prospecting, electrical, 2–3, 12;
    see also Electrical coring and
    Logging
Prospectors, profile of, 34–39, 82–83
Provost, "Maté," 35
Pure Oil, 83
Puzin, 79
Pyinmana, 88

Rangoon, 86, 88
Reconnaissance drilling, 11
Regis, 11
Research Center, 130
Resistivity method, 8–9, 12–13,
    25, 82, 108
Revue des Elèves, 34
Riboud, Jean, 125, 133–134
Ridgefield, Connecticut, 129, 130,
    133

Rio de Janiero, 89
Roche, 48, 83
Romania, prospecting in, 12–13, 84, 95, 112, 119–120, 122
Royal Dutch Shell Company. *See* Shell Oil Company
Roxana Petroleum Company, 15, 22, 29

Saigon, 89
Sain-Bel, 3
Saint-Gaudens, 113
Saint-Most, 113
Salt domes, discovery of, 17–20, 74
Samarkand, 78
San Fernando, 84
Sarasota, 129
Sauvage, Raymond, 15, 35, 56, 59–60, 81, 87–90
Scheibli, Charles, 13, 36, 112
Schlumberger Company
    American operations, 41–52, 93–103, 105–109
    Asia, prospecting in, 85–91
    first steps, 11–16
    international operations, 33–39
    international structure, 119–125
    new enterprises, 127–131
    origins, 1–9
    South America, prospecting in, 81–85
    Soviet Union, prospecting in, 53–80
    subsurface measurements, 25–31
    war years, 111–119
Schlumberger, Conrad, 1, 2, 6–9, 11, 12, 14, 15, 17, 18–19, 21–24, 28, 29, 36, 42–43, 53–54, 56–63, 66–70, 72–75, 77, 79, 95, 97, 105, 111, 120, 121, 134, 135

Schlumberger Electrical Coring, 102
Schlumberger Electrical Prospecting Company, 102, 103
Schlumberger Electrical Prospecting Methods, 41, 105
Schlumberger, Jean, 7
Schlumberger, Jeanne, 8
Schlumberger, Louise, 1–2, 8
Schlumberger, Marc, 14–16, 31
Schlumberger, Marcel, 5–9, 14, 15, 17, 18–19, 23, 24, 28, 37, 42–43, 45, 53–54, 69, 77–78, 79, 95, 106, 107, 112, 113, 114, 117, 120, 121, 133–135
Schlumberger, Nicolas, 5
Schlumberger Overseas, 124, 130
Schlumberger, Paul, 3–7, 24
Schlumberger, Sylvie, 102
Schlumberger Technical Review, 97
Schlumberger Well Surveying Corporation, 100, 103, 105–109, 114, 116–117, 123, 127, 133, 134
Schneersohn, Boris, 37, 89, 112
Seismology, 12
Seminole, 46–49
Seydoux, René, 112, 125
Shell Oil, 81, 82, 83, 88, 93, 95–96
Silvestre, 89
Singapore, 88
Societé de Prospection Electrique, 8, 53, 67, 122–124
Société de Prospection Géophysique, 94–95
Société Géophysique des Recherches Minières, 94
Sondes, 25, 27, 59
Soumont, 3, 33
Southeast Asia, prospecting in, 86–91
South Vietnam, 89

Soviet Union, prospecting in, 8, 35, 53–60, 61–70, 71–80, 84, 95
Soyuzneft, 79
Spain, 33
Spectrometry, 129
Spontaneous polarization, 3, 50, 74
Spontaneous potential curves, 84
Standard Oil Company, 82, 83, 93, 108, 109
Steaua Romano Company, 12
Sterlitamak, 78
Stoll, 89
Sumatra, 88, 89
Surenco, 124, 130
Surface electrical measurements, 11

Tamu, 90, 91
Telemetry, 129
Telluric, 9
Terek, 76, 77
Texas, prospecting in, 14–16, 21–24 93, 95, 98, 99, 102–103, 105, 106
Tierra del Fuego, 125
Tiflis, 78
Tinkin, 90
Tinsukia, 88
Titusville, 45
Tonkin, 89
Trinidad, 84–85, 89, 95

Tulsa, 45, 50
Tucson, 103

Ufa, 78
United States, prospecting in, 8, 14–16, 27–31, 41–52, 74, 105–109, 114
  "boom" in, 93–103
  disappointment in, 21–24
U.S.S.R., prospecting in. *See* Soviet Union
Uzbekistan, 78

Val-Richer, 51
Van Eck, 15
Van Horst, 22
Venezuela, prospecting in, 8, 35, 81–86, 95, 124
Vespucci, Amerigo, 82
Viry, René, 23

Washington, D.C., 51–52
Washington, George, 51
Well-cementing, 108
West Ranch, 98
"Work-overs," 84
World War II, 88–89, 111–117, 119–129

Yenangyaung, 86, 88, 90, 91
Yugoslavia, 3

Zametov, 62, 67